T0331184

Family Farms and the Conservation of Agrobiodiversity in Cuba

This highly original volume investigates and documents the complex interactions between small family farms and Man and Biosphere Reserves in Cuba.

Covering over two decades of research in agriculture and biodiversity conservation in Cuba, this book provides a unique case study about sustainable agriculture. It shows how the agricultural biodiversity maintained in situ by family farms within those protected areas provides a strategic source of crop genetic resources, including seeds and planting materials, as well as agroecological knowledge and practices. Agricultural practices within and around the Biosphere Reserves have helped to promote local food security through healthier and more diverse food production, while contributing to the conservation of biodiversity and of ecosystems. This book also reports on the adoption of transdisciplinary methods, combining ecological, agronomic, and socio-economic research, along with participatory methods involving farmers in research to document ethnobotanical and farmer knowledge, revealing rich spots of agrobiodiversity maintained in landscapes, seed systems, and nurseries managed by farmers. It covers a range of ecosystems and biocultural landscapes from arid tropics, tropical hillsides and savannas, montane rainforests, and coastal areas. It examines how family farms in diverse Cuban ecosystems use biodiversity, agroecological knowledge, and techniques while sustaining natural and farming landscapes in a scenario of climate change, frequent disasters, and socio-economic and policy changes.

This book will be most suitable for those studying or interested in farming practices, biodiversity conservation, food security, agrobiodiversity, and sustainable development, as well as in Cuban studies.

Urbano Fra Paleo is a retired Professor of Human Geography from the University of Extremadura, Spain. He is a member of the European Science and Technology Group (E-STAG), United Nations Office for Disaster Risk Reduction (UNDRR); Honorary Research Fellow of the Alliance of Bioversity International and CIAT, Senior Research Fellow of the Earth System Governance Project, and Member of the Academia Europaea.

Leonor Castiñeiras is a biologist with a PhD in Agricultural Sciences. She worked as Scientific Researcher leading the Plant Genetic Resources Unit at the Instituto de Investigaciones Fundamentales en Agricultura Tropical Alejandro de Humboldt (INIFAT) in La Habana, Cuba. She has coordinated various international projects in partnership with Bioversity International (formerly IPGRI) on the characterization of genetic resources with particular focus on traditional varieties and their seed systems.

Issues in Agricultural Biodiversity
Series editors: Michael Halewood and Danny Hunter

This series of books is published by Earthscan in association with Bioversity International. The aim of the series is to review the current state of knowledge in topical issues associated with agricultural biodiversity, to identify gaps in our knowledge base, to synthesize lessons learned and to propose future research and development actions. The overall objective is to increase the sustainable use of biodiversity in improving people's well-being and food and nutrition security. The series' scope is all aspects of agricultural biodiversity, ranging from conservation biology of genetic resources through social sciences to policy and legal aspects. It also covers the fields of research, education, communication and coordination, information management and knowledge sharing.

Agrobiodiversity, School Gardens and Healthy Diets
Promoting Biodiversity, Food and Sustainable Nutrition
Edited by Danny Hunter, Emilita Monville Ora, Bessie Burgos, Carmen Nyhria Roel, Blesilda M. Calub, Julian Gonsalves and Nina Lauridsen

Biodiversity, Food and Nutrition
A New Agenda for Sustainable Food Systems
Edited by Danny Hunter, Teresa Borelli and Eliot Gee

Orphan Crops for Sustainable Food and Nutrition Security
Promoting Neglected and Underutilized Species
Edited by Stefano Padulosi, E.D. Israel Oliver King, Danny Hunter and M.S. Swaminathan

Family Farms and the Conservation of Agrobiodiversity in Cuba
Food Security and Nature
Edited by Urbano Fra Paleo and Leonor Castiñeiras

For more information about this series, please visit: www.routledge.com/
Issues-in-Agricultural-Biodiversity/book-series/ECIAB

Family Farms and the Conservation of Agrobiodiversity in Cuba

Food Security and Nature

Edited by
Urbano Fra Paleo and Leonor Castiñeiras

Routledge
Taylor & Francis Group
LONDON AND NEW YORK

earthscan
from Routledge

First published 2024
by Routledge
4 Park Square, Milton Park, Abingdon, Oxon OX14 4RN

and by Routledge
605 Third Avenue, New York, NY 10158

Routledge is an imprint of the Taylor & Francis Group, an informa business

British Library Cataloguing-in-Publication Data
A catalogue record for this book is available from the British Library

ISBN: 978-1-138-73571-2 (hbk)
ISBN: 978-1-032-56164-6 (pbk)
ISBN: 978-1-315-18388-6 (ebk)

DOI: 10.4324/9781315183886

Typeset in Times New Roman
by codeMantra

Contents

About the Editors

Urbano Fra Paleo is a retired Professor of Human Geography from the University of Extremadura, Spain. He is a member of the European Science and Technology Group (E-STAG), United Nations Office for Disaster Risk Reduction (UNDRR), Honorary Research Fellow of the Alliance of Bioversity International and CIAT, Senior Research Fellow of the Earth System Governance Project, and Member of the Academia Europaea. He has been Visiting Professor at various universities and research centers in the United States, Mexico, Costa Rica, Japan, and Czech Republic. He also was Research Associate at the Environment Institute of the University of Denver and Fellow of the American Geographical Society Library, University of Wisconsin-Milwaukee, United States. His research focuses on disaster risk governance, displaced communities by disasters, and food security.

Leonor Castiñeiras is a biologist with a PhD in Agricultural Sciences. She worked as Scientific Researcher leading the Plant Genetic Resources Unit at the Instituto de Investigaciones Fundamentales en Agricultura Tropical Alejandro de Humboldt (INIFAT) in La Habana, Cuba. She has coordinated various international projects in partnership with Bioversity International (formerly IPGRI) on the characterization of genetic resources with particular focus on traditional varieties and their seed systems. The study of agricultural biodiversity and its *in situ* conservation have been the most relevant lines of her research, aimed at the conservation of traditional agrobiodiversity with significance in the livelihoods of farmer communities living in Cuban biosphere reserves (UNESCO Man and the Biosphere Programme). Currently she is a consultant to the Alliance Bioversity International and CIAT in the Biodiversity for Food and Agriculture Unit as an expert in the management of genetic resources.

Contributors

Giraldo Acosta Alcolea
Centro Oriental de Ecosistema y Biodiversidad (Bioeco), CITMA, Santiago de Cuba, Cuba

Noel J. Arozarena Daza
Instituto de Investigaciones Fundamentales en Agricultura Tropical Alejandro de Humboldt (INIFAT), La Habana, Cuba

Damaysa Arzola Delgado
Estación Ecológica Sierra del Rosario, Centro de Investigaciones y Servicios Ambientales, ECOVIDA, CITMA, Pinar del Río, Cuba

Gerardo Begué-Quiala
Unidad Presupuestada de Servicios Ambientales Alejandro de Humboldt (UPSA), Ministerio de Ciencia, Tecnología y Medio Ambiente (CITMA), Guantánamo, Cuba

Nadia Bergamini
Alliance Bioversity International and CIAT, Rome, Italy

Osmani Borrego Fernández
Península de Guanahacabibes Biosphere Reserve Centro de Investigaciones y Servicios Ambientales, ECOVIDA, CITMA, Pinar del Río, Cuba

Celia Cabrera Ibáñez
Instituto de Investigaciones Fundamentales en Agricultura Tropical Alejandro de Humboldt (INIFAT), MINAG, La Habana, Cuba

José Alberto Camejo Lamas
Península de Guanahacabibes Biosphere Reserve Centro de Investigaciones y Servicios Ambientales, ECOVIDA, CITMA, Pinar del Rio, Cuba

Pablo Eyzaguirre
Alliance Bioversity International and CIAT, Rome, Italy

Carlos Gallardo Toirac
Centro Nacional de Áreas Protegidas (CNAP), La Habana, Cuba

Alessandra Giuliani
School for Agricultural, Forest and Food Science (HAFL), Bern University of Applied Sciences, Bern, Switzerland

Alejandro González Álvarez
Instituto de Investigaciones Fundamentales en Agricultura Tropical Alejandro de Humboldt (INIFAT), La Habana, Cuba

Maribel González-Chávez Díaz
Instituto de Investigaciones Fundamentales en Agricultura Tropical Alejandro de Humboldt (INIFAT), La Habana, Cuba

Rey F. Guarat Planche
Unidad Presupuestada de Servicios Ambientales Alejandro de Humboldt (UPSA), Ministerio de Ciencia, Tecnología y Medio Ambiente (CITMA), Guantánamo, Cuba

Gaia Gullotta
Alliance Bioversity International and CIAT, Rome, Italy

José Manuel Guzmán Menéndez
Environmental Agency, Ministry of Science, Technology and Environment (CITMA), La Habana, Cuba

Fidel Hernández Figueroa
Estación Ecológica Sierra del Rosario, Centro de Investigaciones y Servicios Ambientales, ECOVIDA, CITMA, Pinar del Río, Cuba

Toby Hodgkin
Platform for Agrobiodiversity Research, Museo Orto Botanico. Università degli Studi di Roma La Sapienza, Rome, Italy

Madeleine Kaufmann
Gesellschaft für Internationale Zusammenarbeit (GIZ), Bonn, Germany

Parviz Koohafkhan
World Agricultural Heritage Foundation, Rome, Italy

Lázaro Lorenzo Ravelo
Instituto de Investigaciones Fundamentales en Agricultura Tropical Alejandro de Humboldt (INIFAT), La Habana, Cuba

Lázaro Márquez Govea
Península de Guanahacabibes Biosphere Reserve, Centro de Investigaciones y Servicios Ambientales, ECOVIDA, CITMA, Pinar del Río, Cuba

Lázaro Márquez Llauger
Península de Guanahacabibes Biosphere Reserve Centro de Investigaciones y Servicios Ambientales, ECOVIDA, CITMA, Pinar del Río, Cuba

Augusto de Jesús Martínez Zorrilla
Centro Nacional de Áreas Protegidas (CNAP), La Habana, Cuba

Zoraida Mendive
Instituto de Investigaciones Fundamentales en Agricultura Tropical Alejandro de Humboldt (INIFAT), La Habana, Cuba

Dunja Mijatovic
Alliance Bioversity International and CIAT, Rome, Italy

Hayler M. Pérez Trejo
Unidad Presupuestada de Servicios Ambientales Alejandro de Humboldt (UPSA), Ministerio de Ciencia, Tecnología y Medio Ambiente (CITMA), Guantánamo, Cuba

Ivette Perfecto
School for Environment and Sustainability, University of Michigan, Ann Arbor, MI, USA

José Puente Nápoles
Ministerio de la Agricultura (MINAG), La Habana, Cuba

Geovanys Rodríguez Cobas
Unidad Presupuestada de Servicios Ambientales Alejandro de Humboldt (UPSA), Ministerio de Ciencia, Tecnología y Medio Ambiente (CITMA), Guantánamo, Cuba

Enrico Ruzzier
World Biodiversity Association onlus, Verona, Italy

Luis Sáez Tonacca
Department of Agricultural Management. Universidad de Santiago de Chile, Santiago de Chile, Chile

Dalia Salabarría Fernández
Centro Nacional de Áreas Protegidas (CNAP), La Habana, Cuba

Yanisbell Sánchez Rodríguez
Instituto de Investigaciones Fundamentales en Agricultura Tropical Alejandro de Humboldt (INIFAT), La Habana, Cuba

Paola De Santis
Alliance Bioversity International and CIAT, Rome, Italy

Tomás Shagarodsky
Instituto de Investigaciones Fundamentales en Agricultura Tropical Alejandro de Humboldt (INIFAT), La Habana, Cuba

Juan A. Soto Mena
Instituto de Investigaciones Fundamentales en Agricultura Tropical Alejandro de Humboldt (INIFAT), La Habana, Cuba

Alberto Tarraza
Instituto de Investigaciones Fundamentales en Agricultura Tropical Alejandro de Humboldt (INIFAT), La Habana, Cuba

Nicola Tormen
World Biodiversity Association onlus, Verona, Italy

José Augusto Valdés Pérez
Centro Nacional de Áreas Protegidas (CNAP), La Habana, Cuba

John Vandermeer
Department of Ecology and Evolutionary Biology, University of Michigan, Ann Arbor, MI, USA

Michely Vega León
Instituto de Investigaciones Fundamentales en Agricultura Tropical Alejandro de Humboldt (INIFAT), La Habana, Cuba

Jorge Luis Zamora Martín
Estación Ecológica Sierra del Rosario, Centro de Investigaciones y Servicios Ambientales, ECOVIDA, CITMA, Pinar del Río, Cuba

Foreword

Cuba is at the interstices of Meso-American and South American crop domestication. The island's demographic history of indigenous, African, Asian, and European peoples, their foods, and cultures have shaped the diversity of Cuban ecosystems, agriculture and tropical food systems. Cuba has also built a robust nature conservation system and established well-managed land-protected areas that meet the UN global conservation targets.

The struggle for food security has been marked by challenges in the global political economy and climate change which have resulted in a shift from monocrop dependency on global markets and food imports to a more diversified economy based on agroecological production and tourism. For many years, the country has been working towards achieving food security which has led to a greater focus on food sovereignty, as the right of everyone to have access to safe, nutritious, and culturally appropriate food in sufficient quantity and quality to sustain a healthy life with full human dignity.

This book describes two decade-long research on Cuba's agricultural biodiversity, documenting and measuring crop genetic diversity associated with biocultural and ecological diversity. The rich diversity of tropical fruits, legumes, grains, roots and vegetables, and small livestock, scarcely found in commercial agriculture and markets, is still thriving in small family farms within and around Cuba's Man and Biosphere (MAB) Reserves. More in particular, this volume summarizes the most recent research carried out within the framework of the international project *Agrobiodiversity Conservation and Man and the Biosphere Reserves in Cuba: Bridging Managed and Natural Landscapes*, funded by the UN Environment Program and the Global Environmental Fund and implemented by Bioversity International in collaboration with the Instituto Nacional de Investigaciones Fundamentales en Agricultura Tropical Alejandro de Humboldt (INIFAT), aimed at securing the conservation of agrobiodiversity in Biosphere Reserves and mainstreaming it into the wider landscapes. The project activities provided essential biological resources and knowledge for more diversified and sustainable agricultural production systems in Cuba. A multidisciplinary team, including agronomists, conservationists, natural and social scientists, both from national and international research institutes, universities, NGOs, and farmers' associations, was established to work in close collaboration with farmers to source, safeguard, disseminate, and explore the hitherto untapped potential for Cuba's agricultural biodiversity that contributes to food sovereignty, healthy diets based on locally sourced foods, and to understand the negative effects of agriculture into natural environments and how these can be minimized and mitigated.

The work carried out in farmers' fields and home gardens revealed the existence of healthy and diverse agroforestry production systems whose structure mimics natural habitats where managed and natural landscapes are integrated without generating conflicts with wild plant and animal species. The rich crop diversity and its related biocultural knowledge were the basis for increasing incomes of family farmers, through the valorization of products coming from low-input and sustainable agriculture, while at the same time contributing to national food security and nature preservation.

This book describes the working experience of the Cuban Ministry of Agriculture and the Ministry of Science, Technology and Environment who joined forces and prompted the collaboration of agricultural scientists, and protected areas managers and international scientists and the establishment of enduring friendships and close working relationships with the hundreds of family farmers reached by the project. This initiated a unique mutual learning process through national and international exchanges, trainings, and organization of public events to educate and raise awareness on the potential and importance of crop diversity and its contribution to the resilience of production systems.

This book wouldn't have been possible without Dr. Pablo Bernardo Eyzaguirre Gonzales to whom this volume is dedicated. To him we owe the original research idea of the project stemming out of the work he had been carrying out for many years in Cuba on the local crop biodiversity, home gardens, and the role of biodiversity for nutrition and food sovereignty. We, Nadia and Paola, had the privilege to work with Pablo during the project designing phase and the first years of implementation and the honour to complete the work, supported by his example and his teachings that have marked our careers.

Pablo Eyzaguirre was a pioneer and a visionary who initiated innovative research linking nutrition, plant genetic resources, and rural livelihoods across the world in support of the needs and rights of local communities and national programs. His knowledge and expertise were recognized and respected by all, high-level scientists as well as farmers in both developed and developing countries. His passion and commitment represented a traction for many colleagues and young scientists for whom he was an example of dedication, integrity, and commitment.

Pablo would have been delighted to see the book completed and published. We thank the editors for helping us document this "journey" full of efforts, challenges, and successes that we have faced together with all the national and international participating partners.

Nadia Bergamini and Paola De Santis,
Alliance Bioversity International and CIAT. Rome, Italy

Introduction

Urbano Fra Paleo and Leonor Castiñeiras

Continuing with practices inherited from previous generations, small farmers in distant areas, far from urban centres, such as in certain sections of eastern and western Cuba have ensured a sustainable livelihood and continuous food supply. Indeed, over time they have spontaneously preserved an agricultural biodiversity which includes common species as well as other unique species and varieties. Only following contact with researchers interested in acquiring knowledge regarding these practices and varieties have farmers become aware of the added value of something previously routine in their lives and considered of no appreciable value for the rest of society.

Since the 1990s, inventories of traditional cultivated species and varieties have been conducted and paired with studies of the socioeconomic conditions of families and communities in relation to the diversity present at each site. These studies identified native cultivated varieties thought to be extinct and others that were unknown, as well as examined the management practices of farmers. For example, the *frijol caballero* bean (*Phaseolus lunatus*) is not a commercial species in Cuba. The variability of this species on the island has been maintained as representing an additional source of vegetable protein in the local diet which has made the conservation of up to three cultigroups of the species possible. Another example is purple corn (*Zea mays*), conserved in eastern Cuba and previously unknown in central and western Cuba. A unique case is the conservation of some varieties of chilli peppers and peppers (*Capsicum annuum-chinense-frutescens* complex) on a single farm, or the traditional cultivars *cachuchita picante* and *arroz con pollo* obtained through selection and identified on only two farms in the western region.

Women farmers have historically played a significant role in the conservation of genetic variability on farms as primary actors in food preparation for the family, as well as selecting species and crop varieties according to taste and culinary qualities to be cultivated on the farm. Local female farmers are also responsible for the cultivation of medicinal plants, as access to health service is problematic on these isolated farms. Furthermore, women are also responsible for the conservation of several ornamental plants through cultivars that produce high-quality flowers in their home gardens.

Farmers adopt agroecological practices including the use of farm-produced organic fertilisers, crop rotation, and do not use chemical products that are

DOI: 10.4324/9781315183886-1

potentially harmful for the environment which all contribute to reducing production costs. Such sustainable agricultural practices and the geographic isolation of these areas have not only mitigated the genetic erosion of agricultural diversity but have also favoured the conservation of the ecosystems in which agriculture is practised. Farmers know equally the local agricultural species and their varieties as well as the natural diversity that is an integral part of their environment since they make regular use of both. It is also true that the low population density, reduced agricultural activity and the difficult access to these areas as well as the rugged terrain are also factors contributing to the low transformation of natural areas. Agricultural practices and the memory of their use, however, are mainly conserved by older generations. This constitutes a challenge for the conservation of biodiversity due to processes such as rural flight, depopulation and the abandonment of agriculture, which are new threats to the conservation of traditional agricultural biodiversity.

Other factors, however, favour its conservation. First, the economic crisis faced by Cuba, which has reduced the food security of the urban population in particular but has also had other side effects. In recent years, existing marketing channels have opened up even further following the changes in legislation permitting the sale of agricultural products of small farmers, thus facilitating the entry of these varieties into the market. The opening of new marketing channels to meet the demand of tourism is particularly promising as this sector is more dynamic in both financial and marketing terms. These production volumes, however, are modest owing to the small number of farmers involved and the low productivity of this system. Consequently, it cannot be considered as a solution for national food security, even if it can provide a valuable contribution in certain areas. The same can be said for the demand from the tourism sector since large hotels require a continuous and homogeneous supply that cannot be met by the production system of small farmers, even when quality assurance systems are applied. Undoubtedly, the lesser tourism sector, which is more local, closer, and direct is more suited to this method of production. Similarly, settlements close to biosphere reserves have benefited from the continuous supply of high quality and organic food produced using traditional biodiversity practices on small farms.

Second, the increased income of farmers generated from the marketing of their products mitigates the abandonment of agriculture and facilitates a more stable rural population, and in particular, helps lessen the loss of these agricultural practices. It is noted that this income in the form of cash provides farmers with cash flow that can be used to improve their home buildings (purchase of building materials and furniture), and their farms (fencing materials to delimit the plots, purchase of traction animals, water storage containers and farming tools), to give some examples.

Large areas of its territory are in a good state of conservation and Cuba has adopted policies to conserve its natural values which over time has led to the protection of large sections and the diversity of ecosystems. Regarding conservation, the biosphere reserve category is worthy of mention since these are areas that seek to acknowledge the compatibility between conservation and the continuity of human activity. Farms linked to these areas are examples of traditional food production systems that guarantee the food security of the farmers and local communities.

Chapter 1 describes the natural biodiversity, ecosystems, and landscapes of the Cuban Archipelago, highlighting that due to its insularity, geology and soil variety, there is a high level of endemism, particularly in the terrestrial biota. Particularly notable are those systems developed on ferritic and serpentine substrates that have the greatest level of endemism in the country.

Diversity guarantees the availability of varied genetic resources and environmental services. Thus, Chapter 2 describes Cuba's protected areas which are managed through the National System of Protected Areas (SNAP in Spanish) not only to protect their biological diversity and the ecosystem services they provide, but also to favour the connection between fragments of ecosystems in natural and managed landscapes. In this context, biosphere reserves acknowledge and promote the maintenance of the equilibrium between human activity and the conservation of natural ecosystems. This chapter also addresses the complex environmental challenges associated with economic development and the risks arising from climate change.

UNESCO's Man and the Biosphere (MaB) Programme aims to achieve a sustainable balance between the conservation of biological diversity – particularly agricultural agrobiodiversity –economic development and cultural heritage. Consequently, the following four chapters describe the ecological values and social dimensions of four different biosphere reserves in Cuba.

Chapter 3 provides a description of the Sierra del Rosario Biosphere Reserve, located in the easternmost part of the Guaniguanico Mountain Range in western Cuba. In addition to addressing biodiversity, it indicates how the process of commercial timber extraction has negatively impacted forest cover in the region. After being declared a Biosphere Reserve, the villagers participated in restoring the forests and improving its environmental quality. Currently, the management plan attributes special emphasis on conservation and the recovery of ecosystem services, as well as the integration of tourism and agricultural activities.

Chapter 4 presents an analysis of the Cuchillas del Toa Biosphere Reserve in the eastern region of Cuba. The core area is the Alejandro de Humboldt National Park, with the highest level of endemism in the country, both in flora and fauna. There are 116 human settlements in the Reserve, well distanced from each other and with very few inhabitants; the agricultural biodiversity associated with the culture of these isolated communities is of inestimable value as it exists only in the eastern region of Cuba. The greatest threat to this reserve is the mining of nickel and other minerals, which has caused severe soil erosion in some areas.

The unique nature of the Guanahacabibes Peninsula Biosphere Reserve is described in Chapter 5. This area is the only Antillean island territory with coasts on both the Caribbean Sea and Gulf of Mexico. This distinctive feature is reflected in the composition of the fauna and, in particular, of the terrestrial flora, as well as in the existence of unique habitats, ecosystems, and landscapes. Many natural species are used by human communities, while agrobiodiversity is highly adapted to the poor soils of the area. Consequently, it is important that conservation and management programmes for agricultural biodiversity resources continue to increase with the growing involvement of local community members in the maintenance and conservation activities. Following this, there has been an increase in the adoption of sustainable agricultural production methods, as demonstrated by the organic certification of Guanahacabibes honey.

Chapter 6 focuses on the Baconao Biosphere Reserve, also in eastern Cuba. This area is characterised by the ecological value of its flora and fauna, as well as by the high level of endemism of some groups, including amphibians, terrestrial molluscs and reptiles and the presence of some exclusive species. The area is home to 52 settlements and most of their agricultural production is consumed within these communities. The existing threats identified include inadequate plantation management, timber extraction, and mining activities that require adaptation measures.

Chapter 7 discusses how biodiversity, from the farmers' point of view – both on-farm and in the environment – is a source of ecosystem services. It also illustrates how agricultural practices bring benefits by favouring a high-quality agricultural matrix through which wildlife can migrate. This chapter describes how the multidisciplinary approach adopted by the COBARB Project proved to be an effective tool to assess the impact of traditional Cuban farms linked to biosphere reserves on sustainability and biodiversity.

The Instituto de Investigaciones Fundamentales en Agricultura Tropical Alejandro de Humboldt (INIFAT) has been working on the *in situ* conservation of traditional plant genetic resources in different regions of Cuba for more than 25 years. Chapter 8 provides an overview of the results of these projects which constituted the foundation of the COBARB project. These results bolstered the recognition of the role of farmers in the conservation of Cuban plant genetic resources for food and agriculture and, consequently, increased their self-esteem.

A first case study, carried out in the Sierra del Rosario Biosphere Reserve and presented in Chapter 9, identifies the factors that define the socioecological production system, which include aspects such as natural and cultivated biodiversity, agricultural practices, environmental quality and socio-cultural factors. It was observed that the natural state of these farms facilitates the movement of organisms between patches, especially coffee plantations in mountain areas which are semi-natural farming systems and well-integrated into the natural environment and, due to their multi-layered vegetation, contribute substantially to the structural connectivity of the landscape.

Chapter 10 shows how traditional family farming and associated agrobiodiversity constitute a productive type of agroecosystem, capable of ensuring both food self-sufficiency and a means of support and income in addition to ecosystem services which can be measured through indicators of social and ecological resilience. It demonstrates that Cuba's landscapes form both a cultural and natural heritage. The level of education of Cuban farmers facilitates an elevated capacity for learning, innovation, adaptation, self-organisation and self-sufficiency, which favours resilience. This chapter shows how the agroecosystems within two biosphere reserves are immersed in a high-quality landscape matrix.

The results of a second case study are illustrated in Chapter 11 which analyses those factors that influence the conservation and management of agrobiodiversity on farms. It shows how agrobiodiversity-rich landscapes are the result of complex agricultural systems that have developed in response to unique physical conditions combined with the cultural and social dimensions of farming communities. It additionally notes how the poor road infrastructure and difficult access renders most communities isolated, representing a challenge for marketing activities and how, because of it, these farms become living and innovative laboratories for agroecology and the transfer of knowledge.

The area closest to the farm home – known as *conuco* – has a higher level of agricultural biodiversity whereas this gradually decreases with distance, thus emulating the stratification of the tropical forest. Chapter 12 reviews the historical role of the *conuco* and its contribution to traditional agrobiodiversity in different production systems (livestock, tobacco, sugarcane, coffee) and its development from colonial times to the present. It also analyses how the Globally Important Agricultural Heritage Systems (GIAHS) approach can be applied to the traditional Cuban *conuco*, in terms of traditional biodiversity, sustainable use and its link to the culture of rural communities. This chapter also highlights how *conucos* associated with coffee cultivation are characterised by a rich local knowledge and ethnographic features, due to the influence of Afro-Cuban religions in the management of their biological resources.

The system of urban, suburban and family agriculture began with the use of commercial varieties of vegetables and when it spread throughout the country, it incorporated other crops (fruit trees, *viandas*, medicinal plants) from the farms. Chapter 13 shows how this system has also been able to absorb large numbers of people entering the farming sector as an opportunity for employment following the cessation of other economic activities. Due to its profound roots in agrarian and agroecological management practices, this system played a vital role in the urgent actions undertaken to tackle Cuba's food crisis. Currently it has reached a high level of organisation and consolidation, demonstrating its contribution to food security as one of the pillars to achieve food sovereignty in the country. Some farmers in the Biosphere Reserves have joined the system to alternatively market traditional agricultural products from their farms.

Notably, it was possible to considerably reduce the *in situ* loss of farm products – especially fruit – by creating links between traditional producers and local markets. Chapter 14 illustrates the structure of the marketing system for agricultural products in Cuba and also outlines the challenge faced by local governments to continue promoting these initiatives. This chapter also identifies opportunities for the creation of value chains and looks at participatory certification systems based on organic agriculture related legislation as an opportunity to widen the market.

Along the same lines, Chapter 15 focuses on a basket of food products and services, based on agrobiodiversity, from those Cuban Biosphere Reserves that have marketing potential. The combination of diverse tourism markets and services grants farmers a range of product choices with recognisable geographic value.

The final chapter gives the reader insight into the perception and interpretation of farmers as observers of the natural space they inhabit as well as of the biodiversity and functions of wild plants and animals, including microorganisms. We understand how they are aware of the opportunities offered by ecosystem services to increase agricultural productivity, and how diversity is an opportunity as a market strategy. Farmers have adapted management practices to the type of farm and its geographical location. At the same time, both the workshops and the seed and agricultural biodiversity fairs organised by the COBARB project have provided a platform for farming families to meet and exchange knowledge, which they then implement on their farms.

Traditional agricultural biodiversity conserved and managed by smallholders in their farming systems continues to provide stability and adaptability to rural communities, forming part of the family livelihood strategy. Conserving and maintaining this traditional agricultural biodiversity through its utilisation is a powerful course of action to advance the country's food security and sovereignty, while protecting natural biodiversity and landscapes, and securing resilience to environmental changes resulting from climate change.

1 The Cuban Archipelago

Biological diversity, ecosystems and landscapes

José Manuel Guzmán Menéndez

The greatest biological diversity in the Antilles is found in the Cuban Archipelago due to the wide range of species and the degree of endemism, making the islands a biogeographic province within the region. Endemism in Cuban terrestrial biodiversity is mostly concentrated in the eastern and western regions of the main island, particularly in mountainous terrain, in areas with toxic and poor soils such as serpentine and siliceous sand, in arid zones (south-eastern coast), and in regions of high rainfall (east and northeast).

Cuba is no stranger to the numerous and complex environmental problems that development has generated in the interaction between society and nature. The source of these problems can be attributed to the very specific economic and social development of the country – from the Colonial period to the present day – that led to a loss of biodiversity, and which is now the target for actions of the *Estrategia Ambiental Nacional* (Delgado, 1999; CITMA, 2016). Progressive stages in socioeconomic development have generated greater awareness of environmental values by Cuban society for the conservation and utilisation of biological diversity, especially following the Rio Summit (UN, 1992). Only by maintaining a good state of conservation of the biota in diverse ecosystems can ecosystem services, defined as basic elements to achieve the well-being of humankind by the Millennium Ecosystem Assessment (2005), be guaranteed.

The *Estudio Nacional sobre la Diversidad Biológica en la República de Cuba* (Vales et al., 1998) attributes the cause of habitat transformation primarily to deforestation, but also to the effect of other risk factors associated to global change and natural hazards which are the main cause of habitat fragmentation. Ultimately, the loss of biological diversity is further exacerbated by climate change and other anthropic actions (Suárez et al., 1999; Centella et al., 2001; CITMA, 2009). Forest clearing in preparation for the 1970 sugar harvest further reduced forest cover to 15.0% (Del Risco, 1995), its lowest level ever. Reforestation started in 1968 in the western region of the country (Las Terrazas, Pinar del Río Province) as part of a socioeconomic development project. Subsequently, and with the same objectives, other projects were implemented which increased the forested territory of the island to 25.26% in 2017. These actions favoured the conservation of other natural resources, especially those ecosystems and original natural landscapes fundamental in maintaining geoecological conditions in the Cuban Archipelago (Fernández

DOI: 10.4324/9781315183886-2

and Pérez, 2009) as well as representing an environmental achievement for the country. On-going reforestation is primarily to satisfy the demands of the country's economy for different types of wood, but woodlots are also planted near water bodies, drainage basins, and in steep mountainous areas to protect water and soil. Undoubtedly, this activity still has room for improvement such as in the selection of species and reforestation areas, levels of seedling survival, suitability of plantation, as well as improving structure and species diversity which are far from optimal.

The diversity and variability of the Cuban biota, with 34,767 native and 732 introduced species (Mancina and Cruz Flores, 2017), is associated with the great variety of landscapes and ecosystems derived from its diverse geology and geomorphology which includes mountainous systems in the eastern and western regions, an extensive coastal area, plains, and hills. The plains – found throughout the country – are areas that have undergone most transformation through socioeconomic development. The north coast of Cuba is 3,209 km long and the south coast 2,537 km, while the marine platform covers an area of about 67,831 km², which highlights the value of the coastal and marine ecosystems hosting Cuban biota (Menéndez, 2013).

Terrestrial biota

The Cuban terrestrial biota includes more than 25,733 known native taxa, with a 43.0% degree of endemism. There are 6,643 species of flora, of which Gymnosperms (conifers) with 63.2% endemism and Angiosperms (flowering plants) with 52.6% endemism are notable.

There are 12,860 species of fauna with invertebrates registering the greatest numbers and 40.0% endemism; molluscs with 66.3% endemism; and arachnids with 46.2%. There are 280 species of birds (the largest group of vertebrates in the country) although the endemism of this group is only 10.0%. The largest number of vertebrate endemics are amphibians with 96.6%, while reptile endemism is 87.3%. Overall, the highest percentage of endemism is found in flowering plants, molluscs, insects, amphibia, and reptiles.

Many of these species, however, are endangered. Indeed, 70.5% of Cuban species of flora have been categorised as endangered, representing 14.0% of the native flora and 75.0% of endemic flora taxa of Cuba are in the endangered category. The largest number of endemic taxa are in areas with a diversity of substrates such as forests in mountainous areas (especially rainy and pine forests), coastal scrub and other scrub formations commonly found on serpentine soils.

Invasive alien species (IAS) alter the state and functioning of ecosystems and landscapes and represent a threat to the conservation of biological diversity. IAS are being studied with standardised methods that allow the identification, management, and monitoring of the national inventory of this group as well as the evaluation of the ecosystems and areas with the highest level of invasion. In terrestrial ecosystems, various invasive exotic plant species with the highest level of ecological plasticity have been identified: marabú (*Dichrostachys cinerea*), tropical almond (*Terminalia catappa*), aroma (*Vachellia farnesiana*), casuarina (*Casuarina equisetifolia*), aroma blanca (*Leucaena leucocephala*), rose apple (*Syzygium jambos*), caña brava (*Bambusa vulgaris*), espatodea (*Spathodea campanulata*), niaulí (*Melaleuca quinquenervia*), Madras thorn (*Pithecellobium dulce*), and others (Regalado et al., 2012; Oviedo and González-Oliva, 2015). Very harmful invasive mammal species have also been identified, such as the black rat (*Rattus rattus*), house mouse or guayabito (*Mus musculus*), cat (*Felix catus*), small Indian mongoose (*Herpestes auropuntatus*), jíbaro dog (*Canis lupus*) and the wild boar (*Sus scrofa*) (Borroto-Páez and Mancina, 2011).

Aquatic biota

The marine flora and fauna of Cuba show a greater diversity of species than other Caribbean islands. This seems to be determined by a relatively extensive marine platform, which favours the retention of local-species larvae as well as the recruitment and settlement of oceanic larvae from remoter regions.

A distinctive feature of marine ecosystems is their low level of endemism. There is significant connectivity and multiple relationships among the aquatic biota and, consequently, biogeographic provinces tend to have few unique species. The number of known marine species is less than that of terrestrial species and their taxic diversity (higher taxa) is greater than in land-based species. Taking into consideration the connection of the Cuban marine platform with other islands of the Greater Caribbean, we can expect the flora and fauna to show a low level of endemism. The richness of species, variety of habitats, and state of conservation, however, characterise this region as one of the most biologically diverse in the Western Hemisphere. Marine invertebrate species registered in Cuba are in excess of 5,700 and chordates total more than 1,060 (mainly fish). In addition, some 7,300 species of microorganisms and marine flora are known. It is estimated that the number of possible species in the marine ecosystem could exceed 10,500. Based on this estimate, it is inferred that at least 30.0% of marine species of flora and fauna of Cuba are yet to be discovered. This percentage could be much greater for microorganisms and deep-sea fauna, which – due to their lower accessibility – have been much less studied.

Some invasive exotic species in Cuba's aquatic ecosystems have been identified, such as the water hyacinth (*Eichhornia crassipes*), Waterthyme (*Hydrilla verticillata*), elodea (*Elodea canadensis* and *Egeria densa*), and lechuguilla (*Pistia stratiotes*). Very adaptive exotic fish species are found in freshwater ecosystems, such as the African catfish (*Clarias gariepinus*), among others (De la Rosa and Campbell, 2008). In addition, it is documented that marine ecosystems have been invaded by the lionfish (*Pterois antennata*).

Ecosystems and landscapes

The Cuban archipelago has a great variety of ecosystems and terrestrial landscapes. The areas conserving the principal natural biotic resources, with highly natural and representative ecosystems and landscapes, count for 14.0% of the national territory. These areas have undergone less transformation due to their difficult access as they are located primarily in mountainous ranges, swamps, and wetlands.

The original forest cover of Cuba – estimated at between 70.0% and 80.0% (Del Risco, 1995) – is represented mainly by semi-deciduous and evergreen forests. It is also estimated that 90.0% of original forests still existed up until 1812. Forests are mainly represented by humid tropical formations, the boreal limit for this type of forest due to the latitude of Cuba. Forest formations in Cuba range from tropical rainforests (jungles) and cloud or mossy forests to evergreen forests, wetlands, and mangrove forests, as well as semi-deciduous and pine forests. Scrublands develop as xerophytic formations along the coast and inland areas, with semi-desert coastal and subcoastal scrub, as well as other types that develop on ferritic and serpentinic substrates which host the largest number of endemisms. A particular type of montane scrub (*subpáramo*) is found in Pico Turquino (Sierra Maestra), the highest peak in Cuba at 1,974 metres above sea level. Herbaceous plant communities are represented by different types of savannahs, mainly edaphic and semi-anthropic, freshwater, and related communities present in streams and permanently or periodically flooded areas, as well as by communities of halophytes.

Vegetation complexes are groups of plant communities in which different physiognomies and even different types of vegetation concur in the same landscape unit, such as isolated steep-sided residual hill or *mogotes* (conical-karst formations), as well as rocky and sandy coastal complexes present in Cuba. Secondary vegetation has an interesting floristic and structural complexity, determined by different successional stages and anthropogenic factors, and includes forests, scrub, and herbaceous communities. Ruderal and segetal vegetation are associated with human activity and found particularly in inhabited and cultivated areas.

Coral reefs occupy more than 98.0% of the 3,215 km of the marine platform which is bordered by frontal reefs and ridges as part of barriers. Approximately 1,440 km of reefs are found in the north and 1,675 km in the south of the main island. In addition, many reefs are scattered over wide areas but within the island shelf, two-thirds of which are at risk. The rich muddy seabed is highly productive and is an important source of fishery resources. This typology of seabed is commonly found in extensive estuarine areas, coastal lagoons and mangrove lined shores. Seagrass meadows cover more than 50.0% of the seabed of the Cuban platform (Martínez Daranas, 2010). These are the main entry points for matter and energy that guarantee biological and fishing productivity on the platform, part of which is exported to the reefs and ocean, thus constituting an important ecological reserve. Mangroves cover approximately 4.8% of the national territory and accounts for 26.0% of the country's total forested area. Most of the Cuban Archipelago coast is bordered by mangroves, as are other swampy areas, coastal lagoons, and estuaries (Menéndez and Guzmán, 2006), which indicates their importance. The main marine biotopes in Cuba are coral reefs, uncolonised hard-bottom habitats, non-reef hard-bottom habitats (inland waters), unconsolidated sediments (sand and mud), submerged vegetation (seagrasses and macroalgae), mangroves, coastal lagoons and estuaries, and low rocky coasts either with cliffs or beaches.

The outer beaches of Cuba are formed mostly by biogenic and oolitic-biogenic materials where much natural erosion and alterations in the sedimentary equilibria of coastal systems are observed, mainly affecting seagrass meadows and coral reefs. Inner beaches are predominantly composed of biogenic marine and terrigenous sediments, and often a mixture of both. In general, erosion rates are of moderate intensity in most beaches in the archipelago, with rates no greater than 1.2 m/year (Tristá Barrera, 2003; Rodríguez et al., 2009), although some higher values may be found.

The ecosystems and natural terrestrial landscapes of the Cuban Archipelago were evaluated for fragmentation of vegetation cover to calculate its representativeness and state of conservation (Capote et al., 2007, 2011) by examining 750 polygons or patches. Each vegetation polygon was associated with the capacity of the habitat to sustain the vital processes of biological diversity. The level of fragmentation of the vegetation cover ranges from medium to high, with vegetation fragments of up to 1,000 km². For the wetland category, mangrove forests and swampy grasslands have the most extensive patches of natural vegetation whereas agricultural areas can have patches of more than 1,000 km². Finally, secondary vegetation is characterised by medium to high fragmentation, a value similar to natural vegetation.

Seagrass meadows and mangrove forests are the most common wetlands by area (Martínez Daranas et al., 2009; Menéndez, 2013) and are associated with related ecosystems such as swamp grassland, swamp forest, and semi-deciduous forest with fluctuating humidity. In addition, mesophyll tropical forests (submontane, swamp, coastal and subcoastal mesophyll evergreen forests, semi-deciduous and fluctuating-humidity mesophyll forests, and pine forests) have lost more than 50.0% of their potential area, primarily due to favourable conditions for urbanisation and agriculture. In general, the loss of landscape values found in urban, rural, and natural areas of the country is attributable to construction or unauthorised activities, or lack of any preliminary environmental impact assessment and, additionally, to the non-compliance with indications or recommendations arising from any assessment.

Biological diversity in the face of climate change

Cuba has documented the impact, vulnerabilities and possible adaptation pathways of Cuban biodiversity to climate change. These findings are collated in two *Comunicaciones Nacionales a la Convención sobre Cambio Climático* (National Communications to the Convention on Climate Change) (Centella et al., 2001). Subsequently, the impact of climate change on terrestrial biological diversity in the insular Caribbean was reviewed in order to develop a regional research agenda and help identify capacity-building requirements for its implementation (Planos et al., 2013).

The impacts of climate change that have been identified produce direct and indirect effects on biodiversity as well as on goods and ecosystem services and other associated resources such as water, agriculture, and human settlements. This reveals the importance of the coastal areas for the Cuban archipelago. The main impacts of climate change on biodiversity identified (Fernández and Pérez, 2009) are:

- Transformation, reduction or disappearance of species, including physiological and biochemical modifications due to stress processes;
- Effect on coral reefs due to thermal intolerance;
- Reduced catches and fishing yields due to the increase in sea temperature;
- Disappearance of wetland ecosystems due to rising sea levels and changes associated with the hydrogeological regime; and
- Impact on biogeographical zones of the north of the eastern region and, in particular, on endemic plant species according to the biological aridity index model.

The coastal zone – including coastal plains and the shallow waters of the insular platform – shows an increasing level of assimilation and socioeconomic activity. The main impacts of climate change on the coastal zone derive from the rising sea level, causing a potential regression of the coastline and ensuing loss of territory of the main island and its cays with a widespread impact on coral reefs and associated biota due to rising temperatures.

The exchange of water between the shelf and the ocean is also exacerbated as depths increase as well as by changes in the physical-geographical, hydrographic, and hydroclimatic characteristics and an increased risk of coastal flooding due to cyclonic upwelling. The latter has been identified as the greatest hydrometeorological hazard for the Cuban coastline (Capote et al., 2011).

The National System of Protected Areas of the Republic of Cuba

Centella et al. (2001) consider the National System of Protected Areas (SNAP in the Spanish acronym) as the most important measure undertaken for the preservation of Cuban biological diversity against changes driven by climate change. Starting with a study of the Cuban biodiversity, the National Centre for Protected Areas (CNAP in the Spanish acronym) aims to guarantee its conservation and sustainable use by identifying areas of the greatest ecological, social, historical, and cultural relevance. This was not only a priority in the National Environmental Strategy 2006–2010 but also a commitment by the Cuban State as signatory to the Convention on Biological Diversity (CITMA, 2006).

Until 2020, the SNAP included 253 protected areas – of which 91 of national significance – as the best-preserved ecosystems and with greatest natural values in the country. The remaining 162 areas are of local significance. The system covered 19.95% of the national territory in its variants and categories, including the seven existing Special Regions for Sustainable Development, encompassing the five mountain ranges of Cuba (Guaniguanico, Guamuhaya, Bamburanao, Nipe-Sagua-Baracoa and Sierra Maestra), the largest wetland in the insular Caribbean (Ciénaga de Zapata) and the two largest system of cays in the country (the Sabana-Camagüey and Canarreos Archipelagos).

Given that Cuba is an island, the marine-coastal area forms part of a Marine Protected Areas (MPAs) Sub-system within the SNAP. Again, until 2020 there were

108 Marine Protected Areas, of which 21 legally declared and another 13 awaiting final approval.

International recognition of protected areas in the SNAP has been received for the six Biosphere Reserves (Guanahacabibes, Sierra del Rosario, Cuchillas del Toa, Ciénaga de Zapata, Baconao and Buenavista), two World Cultural Heritage Sites (Desembarco del Granma National Park and Alejandro de Humboldt National Park), one World Natural Heritage Site (Viñales National Park), and six Ramsar Sites (Ciénaga de Zapata, Ciénaga de Lanier and the south of Isla de la Juventud, the Río Máximo-Camagüey Wetland, the Wetland of the North of Ciego de Ávila, Buenavista, and the Delta del Cauto Wetland).

The protection of the main natural habitat patches, the identification of buffer zones, the increase of connectivity between fragments, and the need to apply bioregional planning and management were the key criteria used to establish and develop the SNAP in Cuba. These were even more important in the case of the Protected Landscapes/Seascapes (IUCN Category V) and Protected area with sustainable use of natural resources (IUCN Category VI) categories. Mitigation of and adaptation to global changes (including climate change) implies understanding its multilateral, socioeconomic, biophysical and institutional dimensions (Dudley, 2008), so the success of a system of protected areas must be founded on its ability to adapt and respond to the needs of conservation of natural resources as well as the sustainable development in the country.

References

Borroto-Páez, R. and Mancina, C.A. (2011) *Mamíferos en Cuba*, UPC Print, Vaasa, Finland, 271pp

Capote, R.P., Cruz, R.O. and Vantour, A. (2007) 'Fragmentación de Vegetación en el Archipiélago Cubano: Conservación de Diversidad Biológica y Mitigación de Desertificación', *Memorias del Primer Taller Binacional y Regional sobre Desertificación*, IVIC, Caracas, pp.33–36

Capote, R.P., Mitrani, I. and Suárez, A.G. (2011) 'Conservación de la biodiversidad cubana y cambio climático en el Archipiélago Cubano', *Anales de la Academia de Ciencias de Cuba*, vol 1, pp1–25

Centella, A.J., Llanes, Paz L., López, C. and Limia, M. (2001) 'República de Cuba. Primera Comunicación Nacional a la Convención Marco de Naciones Unidas sobre el Cambio Climático', Grupo Nacional de Cambio Climático, 169pp, https://unfccc.int/sites/default/files/resource/Cuba%20INC.pdf, accessed November 5 2022

CITMA (2006) 'Plan de Acción Nacional sobre Diversidad Biológica de la República de Cuba 2006/2010', Ministerio de Ciencia, Tecnología y Medio Ambiente, 29pp, https://www.cbd.int/doc/world/cu/cu-nbsap-v2-es.pdf, accessed November 21 2022

CITMA (2009) 'IV Informe Nacional al Convenio sobre la Diversidad Biológica. República de Cuba', Ministerio de Ciencia, Tecnología y Medio Ambiente, 197pp, https://www.cbd.int/doc/world/cu/cu-nr-04-es.pdf, accessed October 26 2022

CITMA (2016) 'Estrategia Ambiental Nacional 2016/2020', Ministerio de Ciencia, Tecnología y Medio Ambiente, http://repositorio.geotech.cu/jspui/bitstream/1234/2727/1/Estrategia%20Ambiental%20Nacional%202016-2020.pdf, accessed October 15 2022

De la Rosa, D. and Campbell, L. (2008) 'Implications of *Clarias Gariepinus* (African Catfish) propagation in Cuban waters', *Integrated Environmental Assessment and Management*, vol 4, no 4, pp521–522

Del Risco Rodríguez, E. (1995) 'Los bosques de Cuba: Historia y Características', Editorial Cientifíco-Técnica, La Habana, 96pp

Delgado, D.J. (1999) *Cuba Verde en busca de un modelo para la sustentabilidad en el siglo XXI Selección, compilación y edición científica*, Editorial José Martí, La Habana

Dudley, N. (2008) *Directrices para la aplicación de las categorías de gestión de áreas protegidas*, IUCN, Gland, Suiza

Fernández, M.A. and Pérez, R.R. (2009) 'GEO Cuba: Evaluación del medio ambiente cubano', Agencia de Medio Ambiente, 293pp

Mancina, C. and Cruz Flores, D.D. (eds) (2017) 'Diversidad biológica de Cuba: Métodos de inventario, monitoreo y colecciones biológicas', Agencia de Medio Ambiente, La Habana, 502pp

Martínez Daranas, B. (2010) 'Los pastos marinos de Cuba y el cambio climático', in A. Hernández-Zanuy and P.M. Alcolado (eds), *La biodiversidad en ecosistemas marinos y costeros del litoral de Iberoamérica y el cambio climático*, Memorias del Primer Taller de la RED CYTED BIODIVMAR, La Habana, pp43–60

Martínez Daranas, B., Cano Mallo, M. and Clero Alonso, L. (2009) 'Los pastos marinos de Cuba: estado de conservación y manejo', *Serie Oceanológica*, vol 5, pp24–44

Menéndez, L. (2013) 'El ecosistema de manglar en el Archipiélago Cubano: Bases para su Gestión', Doctoral thesis, Universidad de Alicante, 172pp

Menéndez, L. and Guzmán, J.M. (2006) *Ecosistemas de Manglar en el Archipiélago Cubano*, Editorial Academia, La Habana, 471 pp

Millenium Ecosystem Assessment (2005) *Ecosystems and human well-being: Synthesis*, Island Press, Washington, DC

Oviedo, R. and González-Oliva, L. (2015) 'Lista nacional de plantas invasoras y potencialmente invasoras de la República de Cuba', *Bissea* vol 9, no 2, pp1–88

Planos, E., Rivero, R. and Guevara, V. (2013) 'Impacto del cambio climático y medidas de adaptación en Cuba', Agencia de Medio Ambiente, La Habana, 335pp

Regalado, L., González-Oliva, L., Fuentes, I. and Oviedo, R. (2012) 'Las plantas invasoras. Introducción a los conceptos básicos', *Bissea*, vol 6, pp1–21

Rodríguez Paneque, R.A., Córdova García, E. A., Franco Abreu, J.A. and Rueda Rueda, A. (2009) 'La erosión en las playas del litoral de Holguín, Cuba', *Ciencias Holguín*, vol 15, no 1, pp1–20

Suárez, A.G., López, A., Ferras, H., Chamizo, A., Vilamajó, D., Martell, A. and Mojena, E. (1999) 'Biodiversidad y Vida Silvestre', in T. Gutiérrez, A. Centella, M. Limia and M. López (eds), *Impactos del cambio climático y medidas de adaptación en Cuba*, United Nations Environment Programme/INSMET, La Habana, pp164–178

Tristá Barrera, E. (2003) 'Evaluación de los Procesos de Erosión en las Playas Interiores de Cuba', Doctoral thesis. Universidad de la Habana, La Habana, Cuba

UN (1992) 'Conferencia de las Naciones Unidas sobre Medio Ambiente y Desarrollo', Río de Janeiro, Brasil, https://www.un.org/spanish/esa/sustdev/documents/declaracionrio.htm, accessed September 28 2022

Vales, M.A., Álvarez, A., Montes, L. and Ávila, A. (1998) *Estudio nacional sobre la diversidad biológica en la República de Cuba*, CESYTA, Madrid, 488pp

2 The System of National Reserves in Cuba

Conserving biodiversity and ecosystem services

Carlos Gallardo Toirac, Augusto de Jesús Martínez Zorrilla, José Augusto Valdés Pérez and Dalia Salabarría Fernández

The insular Caribbean is one of the most important biodiversity hotspots on the planet due to the high concentration of species and endemism (Mittermeier et al., 2011). For this reason, all strategies for the protection of natural areas in Cuba must have the safeguard of biodiversity and ecosystem services (ES) among its primary objectives.

Various plant formations have been identified in Cuba that can generally be grouped into forests, scrub, herbaceous vegetation, vegetation complexes and secondary vegetation (Capote and Berazaín, 1984). According to the Red List of Cuban Flora, the National System of Protected Areas (SNAP in the Spanish acronym) covers 3,210 plant species (including 1,386 endemisms), of which 1,579 are threatened (González-Torres et al., 2016). Despite the dominant mosaic of agroforestry ecosystem landscapes interspersed with remnants of natural vegetation, all the above formations are present in the SNAP.

The fundamental objective in creating protected areas is to conserve their principal natural values, as well as any associated socio-cultural ones. The conservation of natural heritage by declaring certain areas as protected and securing their sustainable use is a commitment taken by the Republic of Cuba through the signature and ratification of the Convention on Biological Diversity and other international conventions.

Law 201/99 regarding the SNAP defines protected areas as

> certain sections of the national territory, declared in accordance with current legislation and incorporated within the territorial planning, that have ecological, social and historico-cultural relevance for the nation and in some cases internationally, particularly focusing, through an efficient management, on the protection and conservation of biological diversity and their associated natural, historical and cultural resources, with the goals of conservation and sustainable use.

Protected areas should not, however, be interpreted as closed-off spaces in which economic activities are totally limited. Indeed, through financing from international

DOI: 10.4324/9781315183886-3

projects, the SNAP undertakes actions that aim to increase food production and reduce the environmental impact of economic activities. Examples of these initiatives in National Parks include the development of rabbit farming and stable-based goat breeding in the Ciénaga de Zapata National Park, the purchase and installation of *casas de cultivo protegido*[1] for vegetable cultivation in the Ciénaga de Zapata and Pico Cristal National Parks, purchase of water pumping equipment, tools and farm implements for the communities of La Melba, Ojito de Agua and other small settlements in the Alejandro de Humboldt National Park.

Protected areas are divided into different management categories according to the permitted level of human intervention. These vary from Natural Reserves, National Parks, Ecological Reserves, Outstanding Natural Elements, Fauna Refuges, Managed Floristic Reserves, Protected Natural Landscapes and Protected Areas with Managed Resources (PAMR). This last category allows greater sustainable use of natural resources and services to meet local or national needs.

The SNAP Strategic Plan is the normative document to coordinate activities pertaining to the environmental policy for all protected areas in Cuba. The SNAP has issued three Strategic Plans from 2003 to the present. The third plan (2014–2020) closed with 215 protected areas, 79 of national significance (APSN in the Spanish and PANS in the English acronym) and 136 Protected Areas of local significance (APSL in the Spanish and PALS in the English acronym).

The 215 protected areas represent 21.26% of the total national territory including marine areas under the jurisdiction of SNAP, or 17.91% of land and 26.69% of territorial waters (Table 2.1).

Although the number of PANS is less than that of the PALS, the former cover a larger territory (3,109,979.37 ha) because they are generally larger, and host more complete ecosystems. The PANS cover 82.25% of the SNAP, while the PALS cover 17.75%, with the main natural values concentrated in the former (Table 2.2). The greatest number of areas in the SNAP are the Managed Floristic Reserve, Fauna Refuge and Outstanding Natural Element categories.

Numbers have increased steadily from the creation of the first protected area in 1930 to the present, and reaching 120 in 2008, on the outcome of research and focus on preserving natural values as a goal of the environmental policy of Cuba. In 2021, under the new SNAP strategic plan, 24 new protected areas were approved by the Council of Ministers, increasing the total number of protected areas in the country (Figure 2.1).

Currently managed protected areas cover 3,357,320 ha. The 150 managed protected areas account for 88.79% of the areas included in the SNAP, leaving

Table 2.1 Proportion of SNAP area in relation to the total area of the Cuban archipelago

	Area (ha)	SNAP area (ha)	%
Land area	10,988,410	1,963,661.39	**17.91**
Marine area	6,781,590	1,817616.61	**26.69**
Total	**17,769,991**	**3,781,278.00**	**21.30**

Table 2.2 Protected Areas of the National System of Protected Areas (SNAP) in Cuba by management category and level of significance (national/local) in 2021

Management category	Category UICN	Level of significance		Total
		PANS	PALS	
Natural Reserve (NR)	I	4	0	4
National Park (NP)	II	14	0	14
Ecological Reserve (ER)	II	18	13	31
Outstanding Natural Element (ONE)	III	12	23	35
Managed Floristic Reserve (MFR)	IV	7	34	41
Fauna Refuge (FR)	IV	11	36	47
Protected Landscape/Seascapes (PLS)	V	2	22	24
Protected Areas with Managed Resources (PAMR)	VI	11	8	19
Total		**79**	**136**	**215**

PANS: Protected Areas of National Significance; PALS: Protected Areas of Local Significance.

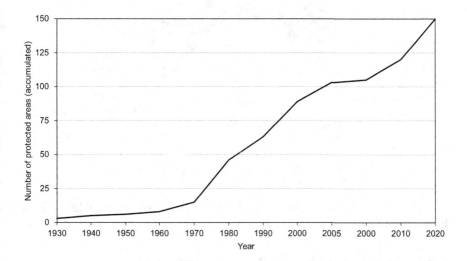

Figure 2.1 Evolution of the number of protected areas administered by the SNAP.

a further 65 identified but unadministered areas. These figures demonstrate that unmanaged protected areas are less extensive and primarily of local importance. The new strategic plan intends to administer protected areas of national importance and a considerable number of protected areas of local significance by delegating their management to Provincial Coordinating Boards of Protected Areas.

The protected areas approved by the Council of Ministers until 2021 total 144, of which 69 of national significance and 75 of local significance, representing 18.53% (Table 2.3) of national territory and 87.08% of the total area of identified protected areas.

Table 2.3 Area (ha) and number of managed protected areas

	Total terrestrial and marine area (ha)	Number of approved protected areas	Area of approved terrestrial and marine protected areas (ha)	Percentage of country area
Total	17,769,991	144	3,292,721.28	18.53%

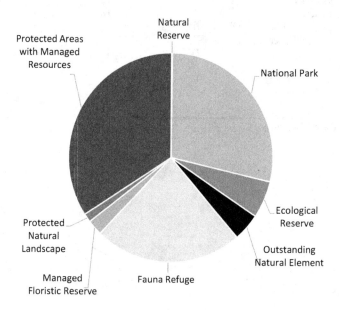

Figure 2.2 Area (ha) of protected areas according to management category.

PAMR is the category with the largest area, in some cases also includes National Parks (Figure 2.2).

Connectivity and biological corridors in protected areas

The concept of biological or ecological corridors implies the existence of connectivity between protected areas and areas with significant biodiversity with the aim of counteracting the fragmentation of habitats. Originally, a biological corridor was conceived as a linear habitat – which noticeably differs from a matrix – connecting two or more fragments of natural habitats (Primack, 2002). This concept, however, has evolved towards a more complete form reaching the level of a mosaic of different types of land use to connect fragments of natural ecosystems across the landscape (Miller et al., 2001; Bennett 2004; Canet-Desanti, 2007).

The SNAP has started to consider connectivity in Cuba as the linkage between terrestrial ecosystems and marine areas functioning as networks which – through currents – facilitate migration, the dispersion of organisms and other natural processes, that contribute with resistance and resilience against climate change. Opportunities to increase connectivity between protected areas within the Cuban archipelago and with other Caribbean islands have been envisaged.

The Cuban archipelago is part of the *Caribbean Biological Corridor*, between Cuba, Haiti and the Dominican Republic, to which Puerto Rico was recently added. This initiative contributes to the integration of conservation actions implemented by the island states, and so favours the preservation of global biodiversity. Its objectives are to reduce the loss of biological diversity in the Caribbean and American Neotropical regions and to streamline communities in harmonious development with nature. The project financed by GEF-UNDP *Application of a regional approach to the management of coastal-marine protected areas in the archipelagos of southern Cuba*, managed to promote connectivity between the selected marine protected areas in this southern region of the country and laid the foundations for advancing the Caribbean Biological Corridor.

In addition, the project *A landscape approach to conserve threatened mountain ecosystems* also financed by GEF-UNDP focuses on the connectivity between protected areas in four mountain ranges, Guamuhaya, Bamburanao, Guaniguanico and Nipe-Sagua-Baracoa. As part of the same project, on-going activities seek to design and implement a pilot corridor in the western part of Cuba. This proposal includes the connection of three core nuclei: the Mil Cumbres PAMR, the Santa Cruz River Canyon (proposed protected area) and the Sierra del Rosario Biosphere Reserve (BR).

Cuban Biosphere Reserves

In 1971, UNESCO launched the Man and the Biosphere Scientific Programme (MaB) which has as one of its objectives the promotion of scientific activities to characterise the biosphere. Research, in conjunction with traditional cultural heritage, aims to formulate strategies for sustainable management of natural resources (CNAP, 2013). Information, experience and good practices are exchanged and shared at a regional and international level through the World Network of Biosphere Reserves. These sites also represent a valuable contribution to the Sustainable Development Goals (SDGs) promoted by the United Nations as contributing to the 2030 Agenda. To date, the National Network of Biosphere Reserves of Cuba – recognised by the UNESCO MaB Programme – has six sites located in different provinces of the country (Figure 2.3).

Sierra del Rosario is an area characterised by low mountains and with a wide range of geological and vegetal formations and highly diverse natural, historical and cultural resources, and a wide variety of ecosystems and habitats. Vegetal formations can be found as evergreen, semi-deciduous and pine forests, *cuabales* (formation growing on poor dry soils, particularly of serpentine origin) and complexes of *mogotes* (isolated steep-sided conical hill of karst origin), with secondary vegetation. As in the whole country, vertebrate fauna is characterised by a scarcity of mammals and a great abundance of reptiles, amphibia and invertebrates (EESR, 2020).

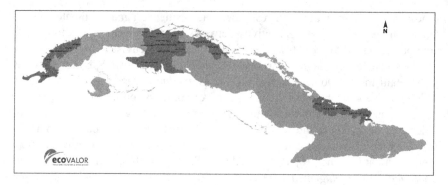

Figure 2.3 Map of the National System of Protected Areas (SNAP) of Cuba with international recognition.

The Guanahacabibes Peninsula has a rich range of ecosystems and a marked marine-coastal predominance and tropical dry forest. Semi-deciduous and evergreen forests, mangroves, coastal and subcoastal xeromorphic scrub and sandy and rocky coastal complexes are the predominant vegetation complexes. The coastline of the southern part of the peninsula is an abrasive karst, interspersed with sandy beaches of varying lengths. The Peninsula hosts a variety of endemic or endangered species, some requiring special protection. The area is particularly significant for birds as it is part of a corridor for more than 50 species during their migration from north to south and vice versa. The marine area protects one of the best-preserved reef ecosystems in the country and one of the most diverse in the Caribbean.

Ciénaga de Zapata is one of the most remarkable territories in Cuba for its extension, integrity and the nature of its ecosystems. The relief is flat and the water table is very close to the surface, making it easily floodable. It forms the largest and most complex karst drainage system in Cuba. It is also considered a phytogeographic district due to the unique composition of its flora and vegetation. Its closest floristic relationship is with the Guanahacabibes Peninsula and the southern part of Isla de la Juventud within the Cuban archipelago, and abroad with southern Florida. Ecologically, this area has very diverse habitats, including swamp grasslands, semi-deciduous forests, xeromorphic coastal forests, a marine area with coral reefs, lagoons, rivers, mangrove forests and flooded caves (Empresa Forestal Integral Ciénaga de Zapata, 2019).

Buenavista has a great geographical diversity which is reflected in the variety of terrestrial, coastal and marine areas of great significance for their elevated ecological sensitivity, high degree of biodiversity and to which the faunal, floristic, speleological and landscape values must also be considered. The Reserve holds significant terrestrial and marine landscapes, including wetlands, coral reefs, cays and islets, forests and caves. Due to their importance for conservation, areas for reproduction, spawning and development of marine species are protected, as well as nesting sites for a large number of aquatic bird species that venture into the coastal mangrove forests and further inland (Pulido et al., 2006).

Cuchillas del Toa is one of the wettest places in Cuba and its irregular relief and extensive forest cover are ideal conditions for the most important and cleanest river waters in the country that form the main water reserves of the island. The area has various plant formations with high biodiversity such as montane rain forest, lowland rain forest, mesophyll semi-deciduous, microphyll evergreen, and pine forests, sub-spiny xeromorphic scrub on serpentine (charrascal), montane or cloudy scrub, secondary scrub, secondary forests, anthropic savannahs and cultural vegetation. The Reserve contains sufficient extension and level of conservation to guarantee the functioning of vital ecological processes and the survival of the local species (Villaverde et al., 2014).

Baconao has two types of coastlines, with abrasive-accumulative cliff and abrasive-denudative patterns. Climate is primarily determined by the relief and the action of the east and northeast winds. Studies to classify plant formations in the area have identified the following: montane rain forest, pine forests, mesophyll evergreen forests, mesophyll and microphyll semi-deciduous forests, gallery forests, *uveral* (coastal vegetation with abundance of uva caleta (*Coccoloba uvifera*), mangrove forest, grasslands, aquatic and *rupicola* vegetation (with predominance of Ericaceae), cloud scrub and coastal and pre-coastal scrub. Of the 22 species of birds endemic to Cuba, 13 species are found in the Reserve (Álvarez et al., 2021).

The BRs are not seen as another category of protected area but are a response to denoting the integration of compatible human values and activities in natural areas. The conservation of the core of the Reserves favours the protection of genetic resources, species of interest, fragile ecosystems and landscapes. Similarly, and with a focus on conservation, economic development for human well-being is sustainably promoted.

Protected areas and the conservation of agricultural diversity

The SNAP plan (2014–2020) adopted the Agrobiodiversity programme – managed by the Alejandro de Humboldt Institute for Fundamental Research in Tropical Agriculture – towards considering protected areas as the core area for species conservation.

Research on the conservation of agricultural diversity in the Sierra del Rosario and Cuchillas del Toa Biosphere Reserves was developed with a multidisciplinary approach aimed primarily to integrate conservation objectives with the agricultural and natural diversity in these sites. Both areas have incorporated participatory approaches to preserve traditional and native plant genetic resources, as a strategy for the sustainable management and use of biodiversity as a whole. An outcome is the wealth of traditional crops that are still cultivated in rural communities today. Generally, traditional varieties have lower yields than improved or commercial varieties, but as the former have been selected and used with little input and resources, they have a greater capacity of adaptation and tolerance to environmental stressors (biotic and abiotic) that characterise the areas where they have developed and so, are invaluable for the sustainability of local production. That is why traditional agricultural systems are reservoirs of diversity that merit more consideration in the country's agricultural and environmental policies.

Due to the agricultural biodiversity and the intraspecific variability in the two reserves, conservation and rural development strategies must be integrated to benefit natural and agricultural diversity and these practices should be extended to other protected areas.

The traditional knowledge of communities, built upon their own agrobiological resources, can contribute towards improving their economic development and enrich the supply of traditional products to local and national agricultural markets. In order to conserve plant genetic resources which are a fundamental part of Cuba's heritage it is essential that local and regional authorities recognise the crucial role of farmers and rural communities in the conservation of plant genetic resources, of national programmes for plant genetic resources and seeds, as well as environmental education programmes.

The integration of rural farmers in the conservation and management plans of Cuba's BRs is essential for the conservation of agricultural heritage, with the purpose of achieving sustainable development based on Cuban native resources found in rural farms although unknown to most of the population. Conserving and promoting farming and consumption by national decision-makers is critical to move towards national food sovereignty (Centro Nacional de Áreas Protegidas, 2013).

Economic valuation of ecosystem goods and services

Many of these ecosystems have been transformed into intensive farming units, which is not their natural function, e.g. forests transformed into agricultural land to produce food. This has an undeniably negative effect on the ecosystems due to the expansion of agricultural and urban frontiers as well as the unsustainable exploitation of its resources and services.

The economic valuation of ES is one of the available tools to measure goods and ES and to be taken into consideration when formulating best practices for decision-making. The GEF/UNDP project *Incorporating multiple environmental considerations and their economic implications in the management of landscapes, forests and productive sectors in Cuba (Ecovalor)* has as its primary goal to promote the generation of multiple environmental benefits using the economic valuation of ecosystem goods and services as a tool for decision-making at different levels. For this to be implemented, work is centred on three components, and scaling the results at the national, provincial and local levels. Starting from an economic valuation analysis of the goods and services of priority ecosystems, such as coral reefs, seagrass meadows, mangrove forests and agroecosystems (agricultural and forestry), financial economic mechanisms are identified to facilitate implementation of good practices. The project involves 5 provinces, 30 municipalities, 15 protected areas, 10 land, water and forest polygons, 7 forest polygons, 3 fishing establishments, 6 tourist centres and 2 hydrocarbon companies (Figure 2.4).

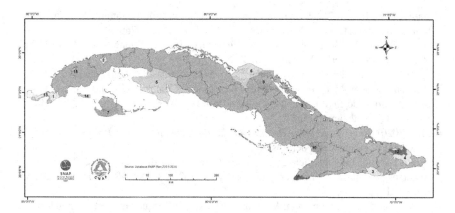

Figure 2.4 Sites (30 municipalities) of intervention of the Ecovalor project. Note: Biosphere Reserves: (1) Península de Guanahacabibes, (2) Sierra del Rosario, (3)
Baconao, (4) Cuchillas del Toa; Biosphere Reserve and Ramsar Sites: (5) Ciénaga de Zapata, (6) Buenavista; Ramsar Sites: (7) Ciénaga de Lanier y Sur de
la Isla de la Juventud, (8) Humedal Río Máximo-Cajuey, (9) Gran Humedal del
Norte de Ciego de Ávila, (10) Humedal Delta del Cauto; Natural World Heritage sites: (11) Parque Nacional Desembarco del Granma, (12) Parque Nacional
Alejandro de Humboldt; Specially Protected Areas and Wild Flora and Fauna in
the Greater Caribbean Region. SPAW Protocol: (13) Parque Nacional Guanahacabibes, (14) Parque Nacional Cayos de San Felipe; Cultural World Heritage
sites: (15) Valle de Viñales.

Early work focused on updating the legal framework, and on internalising
the economic valuation of ecosystem goods and services, as well as the use of
economic-financial instruments, in policies (Land and Environmental Law, Territorial and Urban Planning regulations and Macro programme for Natural Resources
and Environment axis) and regulatory frameworks (Natural Resources and Environment Law, Law of Cultural and Natural Heritage, Decree Law of Protected
Areas, Decree Law of the Coast and its Regulations, Decree on Climate Change).

Note

1 Protected cultivation area: Plot of land covered with screens able to reduce solar radiation by approx. 32.0%, thus facilitating the cultivation of vegetables in periods of intense
 heat. Source: https://www.ecured.cu.

References

Álvarez, L.O., Salmerón, A., Acosta, G., Fagilde, M.C., Álvarez, F., Silot, M., Abad, M.A.,
 Fong, A., Costa, J. and Sanfiel, M. (2021) 'Plan de Manejo para el Paisaje Natural Protegido Gran Piedra. Periodo de manejo 2022-2026', Centro Oriental de Ecosistemas y Biodiversidad (Bioeco)/Empresa Agroforestal Gran Piedra-Baconao, Santiago de Cuba, 366pp
Bennett, A. (2004) 'El papel de los corredores y la conectividad en la conservación de la
 vida silvestre. Programa de conservación de bosques', Serie no 1, UICN, San José, 164pp

Canet-Desanti, L. (2007) 'Herramientas para el diseño, gestión y monitoreo de corredores biológicos en Costa Rica', Master thesis, CATIE, Turrialba, 167pp

Capote, R.P. and R.I. Berazaín (1984) 'Clasificación de las formaciones vegetales de Cuba', *Revista Jardín Botánico Nacional*, vol V, no 2, pp27–75

CNAP (2013) 'Plan del Sistema Nacional de Áreas Protegidas de Cuba: Período 2014-2020 Programa de Agrobiodiversidad y Áreas Protegidas', Centro Nacional de Áreas Protegidas e Instituto de Investigaciones Fundamentales en Agricultura Tropical Alejandro de Humboldt, La Habana, 335pp

EESR (2020) *Plan de Manejo Reserva de la Biosfera Sierra del Rosario 2021-2025*, Estación Ecológica Sierra del Rosario, Las Terrazas, unpublished

Empresa Forestal Integral Ciénaga de Zapata (2019) 'Plan de Manejo Área Protegida de Recursos Manejados Península de Zapata 2020-2025', MINAGRI, Matanzas, unpublished, 210pp

González-Torres, L.R., Palmarola, A., González Oliva, L., Bécquer, E., Testé, E. and Barrios, D. (2016) 'Lista roja de la flora de Cuba'. *Bissea*, vol 10, special no 1, pp1–352

Miller K., Chang, E. and Johnson, N. (2001) *En busca de un enfoque común para el Corredor Biológico Mesoamericano*, World Resources Institute, Washington, DC, 49pp

Mittermeier, R.A., Turner, W.R., Larsen, F.W., Brooks, T.M. and Gascón, C. (2011) 'Global biodiversity conservation: the critical role of hotspots', in F.E. Zachos and J.C. Habel (eds) *Biodiversity hotspots: distribution and protection of conservation priority areas*, Springer, New York, pp3-22

Primack, R.B. (2002) *Essentials of conservation biology*, Sinauer Associates, Sunderland, MA, 698pp

Pulido, E., Martin, G., Blas, S., González, I., González, M.C., Hernández, A. and Pérez, O. (2006) 'Plan de Manejo APRM Reserva de la Biosfera "Buenavista" 2007-2011', Cienfuegos, unpublished, 180pp

Villaverde, R., Begué, G. C., Giraudy, H.M., Pérez, R., Ubals, R., Acebal, Y., Martínez, J., Hernández, N., Correa, P., Sánchez, L. N., Medina, A., Guarat, R. F., Balón, C. Y., Maury, O., Rodríguez, G., Matos, R., Delgado, J. L., Imbert J. R. y., and López, J. B. (2014) Plan de Manejo quinquenio 2014-2020, Parque Nacional Alejandro de Humboldt. Unidad Presupuestada de Servicios Ambientales Alejandro de Humboldt, Delegación Territorial del CITMA, Guantánamo, Cuba, 90 pp.

3 The western mountains

The Sierra del Rosario Biosphere Reserve

Fidel Hernández Figueroa, Jorge Luis Zamora Martín and Damaysa Arzola Delgado

The Sierra del Rosario Biosphere Reserve (RBSR in the Spanish acronym) was the first Biosphere Reserve in Cuba declared by UNESCO in Cuba, on 15 February 1985. It is located in the most easterly part of the Cordillera (range) de Guani-guanico and covers an area of 25,000 ha (250 km²) in the province of Artemisa. The Reserve is currently classified as an Área Protegida de Recursos Manejados (Protected Area with Managed Resources), level 8 in the National System of Protected Areas of the Republic of Cuba (CNAP, 2013). The Reserve is organised around three

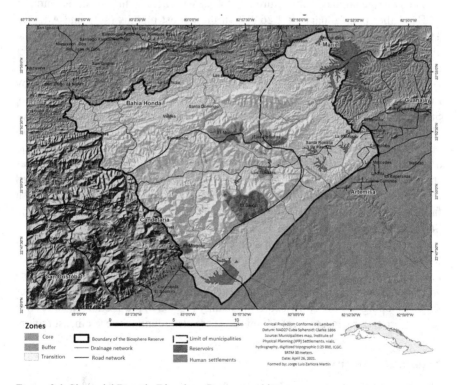

Figure 3.1 Sierra del Rosario Biosphere Reserve and its core zones.

DOI: 10.4324/9781315183886-4

core zones, the Reserva Natural Las Peladas, the Reserva Natural El Mulo, and the Reserva Ecológica El Salón (Figure 3.1) and approximately 8,000 inhabitants are distributed within or in the vicinity of the park in 21 settlements.

Sierra del Rosario is a highly complex tectonic region with numerous and varied overlying thrust belts conferring an allochthonous character to the Mesozoic and lower Palaeocene sequences. Rocks found in the lower Eocene-Quaternary interval and considered autochthonous are deposited transgressively over the oldest folded and faulted zones (Furrazola, 1988).

The varied relief includes plains, highlands and mountains. Towards the north there are slightly declining erosional-denudative plains with hills; to the south, slightly undulating accumulative fluvial-marine plains. Most of the protected area is formed by highlands ranging from 100 to 500 m above sea level and above, the higher altitudes with El Salón peak reaching 565 m. The highlands and mountains are of an erosional denudative type with steep slopes and sharp peaks, though in sectors with a predominantly carbonate structure, slopes can also be vertical with cupola-like peaks, and karstic processes where small caves and canyons are common. In the remaining mountainous sector effusive-sedimentary sediments surface. These mountainous areas have valleys of varying forms depending on the type of rock and sediment. Active and inactive fluvial canyons can be identified in the carbonate material areas and V-shaped valleys in areas characterised by effusive-sedimentary materials.

The average annual temperature is 24.4°C, with 21.3°C in the coldest months of December and January, and 26.8°C in the hottest months (July and August). The minimum temperature recorded was 3.8°C and the maximum, 36.2°C. Annual average precipitation is 2,011.9 mm, the wettest month being June and December the driest month. The average relative humidity is 95%.[1]

River basins follow a north-south pattern, separated by a watershed formed by Las Peladas, El Mulo and Peña Blanca peaks. The San Francisco, San Claudio, Santiago and San Diego rivers flow down the north slope. The San Juan river – with its series of natural basins forming the Baños de San Juan del Norte – flows on the south slope. To the west is the Bayate river with numerous natural basins and a sulphur water spring. The main tributary of the Bayate river is the Manantiales stream which has a waterfall with a 22 m drop. Many small and medium-sized reservoirs are found in the area and constructed to attract leisure activities, fish farming and as a support for agriculture of the region, among which El Palmar and San Juan rivers and the San Francisco dam.

Due to the complex geology and diversity of rocks, several types of soils have formed, mainly fersiallitic, brown and humic (Hernández Jiménez, 1999). Fersiallitic soil is widespread, mainly in the centre of the protected area, in the highlands and mountains. It is also found in small patches in the south and southeast on flat or slightly sloping terrain. Brown soils are mainly located in the south on a large strip running from northeast to west. Humic soils are also found in the southeast region.

Natural biodiversity

The inventory of the Reserve includes 889 organisms, of which 608 are spermatophyte (trees, shrubs and herbaceous plants). The most important families within this group are Poaceae Barnhart with 69 species, Leguminosae Jussieu, nom. cons. with 44 species, Rubiaceae Juss. with 36 species, Asteraceae Bercht. & J. Presl. with 26 species, Euphorbiaceae with 25 species and Cyperaceae and Myrtaceae Juss., with 17 species each. Four species of forestry interest include *Pinus caribaea* Morelet (pino macho), *Samanea saman* (Jacq.) Merr (Cow bean tree), *Talipariti elatum* (Sw.) Fryxell (majagua) and *Tectona grandis* L.F. (teak). There are some endangered endemic species in the area, such as *Acuneanthus tinifolius* (Griseb.) Borhidi, *Ardisia dentata* (A. DC.) Mez, *Gonzalagunia sagraeana* Urb., and *Piscidia havanensis* (Britton & P. Wilson) Urb. & Ekman (Urquiola et al., 2010). Additionally, 28 species of mosses and 55 of liverworts have been collected, this last one includes 34 species belonging to the Lejeuneaceae Cas-Gil family. In the fungi group, there are higher fungi (64 species), hyphomycete (117 species) and 13 species of Endogonaceae. The most important family in this last group is Glomeraceae, with eight species reported. Lastly, there are four genera of lichens, mainly found on the trunks of trees and on the leaves of the arboreal and shrub strata.

Forest formations include evergreen, semi-deciduous, pine forest, xeromorphic scrub and others. Secondary vegetation is reported as becoming dominant in areas of greatest transformation. Evergreen forests at higher altitudes have a good level of conservation. They are found in deep and protected ravines with characteristics similar to the rain forest, at heights between 200 and 400 m above sea level. The dominant stratum reaches an altitude between 20 and 30 m, and in some cases reaching between 35 and 40 m. Many species are found, including *Erythrina poeppigiana* (Walp.) O.F. Cook, *Ficus aurea* Nutt., *Sapium jamaicense* SW., *Matayba apetala* Macf. RDKL, *Pseudolmedia spuria* (SW.) Griseb and *Guarea guidonia* (L.) Sleumer. The well-preserved evergreen forest at mid-altitude has a 10–20 m stratum forming a continuous canopy where some prominent species even reach up to 30 m. The most common species include *Pseudolmedia spuria* (Sw.), *Oxandra lanceolata* (Sw.) Baill., *Matayba apetala* (Macf.) Radlk., *Trophis racemosa* (L.) Urb., *Talipariti elatum* (Sw.) Fryxell (majagua), *Cedrela odorata* M. Roem., *Calophyllum antillanum* (Britt.), *Prunus occidentalis* Sw., *Cinnamomum triplinerve* (R. and P.). Finally, the evergreen forest at lower altitudes has a vegetation similar to the former, although with an arboreal stratum from 3 to 15 metres with some species reaching 20 m tall, with a very dense and irregular canopy. The most abundant species include also *Coccoloba retusa, Matayba apetala, Calophyllum pinetorum* Bisse., *Calophyllum antillanum, Guettarda valenzuelana*, among others (Capote et al., 1988). The semideciduous forest shows two arboreal strata. The upper stratum, between 18 and 20 m, is composed primarily of deciduous trees, while the lower stratum, between 6 and 12 m, is composed primarily of evergreen vegetation (Capote et al., 1988). This vegetation includes, among others, *Ceiba pentandra* (L.), *Ficus crassinervia* Desf. ex Willd., *Cedrela odorata, Andira inermis* (Sw.) HBK, *Bursera simaruba* (L.), *Amyris balsamifera* (L.), *Amyris elemifera* L., *Trichilia hirta* (L.), *Picramnia pentandra* (Sw.), *Adelia ricinella* (L.), *Diospyros crassinervis* (Krug et Urb), *Allophylus cominia* (L.) Sw.

One hundred twenty-six species of birds grouped into 17 orders and 38 families, of which 16 are endemisms, have been reported in the area. Also reported are three of the six genera endemic to Cuba, with the species *Starnoenas cyanocephala*

Linnaeus, 1758 (paloma perdiz), *Teretistris fernandinae* Lembeye, 1850 (chillina), *Xiphidiopicus percussus* Temminck, 1826 (carpintero verde). Other endemic species are *Priotelus temnurus* Temminck, 1825 (tocororo) and *Todus multicolor* Gould, 1837 (cartacuba) (Guerra and Mancina, 2010). Twenty-four species of reptiles have been reported, of which 17 (71%) are endemic, belonging to two orders, one sub-order and 11 families. Various types of lizards from the genus Anolis have been identified: *Anolis luteogularis* (Noble and Hassler, 1935), *Anolis homolechis* (Cope, 1864), *Anolis allogus* (Barbour and Ramsden, 1919), *Anolis sagrei* (Cocteau in Duméril and Bibron, 1837), *Anolis barbatus* (Garrido, 1982), and the *Anolis vermiculatus* (Cocteau in Duméril and Bibron, 1837) (Guerra, 2010). The following snakes have been identified: *Epicrates angulifer* (Bibron, 1843) (majá de Santa María), which is the largest snake species found in Cuba. Individuals of *Cubophis cantherigerus* (Bibron, 1843) and *Antillophis andreai* (Reinhardt and Lütken, 1862) have also been identified.

Nineteen species of amphibians have been registered, with an 84% of endemism, pertaining to one order and four families. Anuran amphibians are abundant in this region, particularly species of the genus *Eleutherodactylus*, such as *Eleutherodactylus zugi* (Schwartz, 1958), *Eleutherodactylu seileenae* (Dunn, 1926) and *Eleutherodactylus symingtoni* (Schwartz, 1957). Also, the species *Eleutherodactylus limbatus* (Cope, 1862) is found, one of the smallest frogs in the world, *Peltophryne fustiger* (Schwartz, 1960), and a toad of the genus *Bufo*, *Osteopilus septentrionalis* (Cope, 1862; Guerra, 2010). Twenty species of freshwater fish were identified, 11 of which are endemisms and 3 local endemisms. The remaining species have been introduced, especially *Clarias gariepinus* (Burchell, 1822) (African catfish), a particularly voracious species that is a threat for native and endemic species (Ponce de León et al., 2010). Mammals are mainly rodents of the Capromyidae family, such as *Capromys pilorides* (Say, 1822) (jutía conga) and *Mysateles prehensilis* (Poeppig, 1824) (jutía carabalí). Yet, the rat *Rattus rattus* (Linnaeus, 1758) and the Javan mongoose *Herpestes javanicus* (Hodgson, 1836) were introduced (Guerra and Mancina, 2010). Bats are an important group of mammals with thirteen species. The most notable are *Pteronotus parnellii* (Gray, 1843), *Artibeus jamaicensis* (Rehn, 1902), *Phyllops falcatus* (Gray, 1839), *Lasiurus pfeifferi* (Gundlach, 1868) and *Phyllonycteris poeyi* (Gundlach, 1868). Studies carried out between 2010 and 2012 in joint expeditions by the Faculty of Biology of the University of La Habana and Ecology and Systematics Institute, reported 26 species from the Lepidoptera order and 49 subspecies of the Hesperiidae (Latreille, 1809), Papilionidae (Latreille, 1802), Lycaenidae (Leach, 1815) and Pieridae (Duponchel, 1835) families, of which 19 species and 11 subspecies are endemic (Núñez Aguila and Barro Cañamero, 2012).

Landscapes

Ten different landscape units have been described, they are formed by two lithostructure depressions, one strongly dissected in serpentinite limestones and other melange, gently dissected undulating over sandstone, limestone and breccias

(González, 1989). The same author describes two types of plains; a slightly inclined and hilly erosive denudation plain which is intensely dissected in limestone, sandstone, serpentine and igneous rocks; and an accumulative fluvial-marine plain, slightly undulating, inclined and dissected in limestones and marls. There is a landscape unit formed by aligned erosive denudation litho-structural hills strongly dissected in sandstone, limestone and schist (González, 1989). Five different units have been described for the more elevated sections of the territory. Highlands with an erosive denudation litho-structure, intensely dissected in limestone and schist (González, 1989). The aligned litho-structural highlands – partly denuded and eroded mogotiphorm – is highly dissected in flysch limestones and metamorphic rocks; the erosive denudation litho-structural highlands is heavily dissected in very large fragments of calcareous breccia and melange serpentinite; the monoclinal horst erosive denudation litho-structural highlands strongly dissected in limestone; and, finally, the anticlinal horst highlands strongly dissected in limestone (González, 1989).

Historical and socioeconomic processes

This history of the territory of the Sierra del Rosario Biosphere Reserve is linked to the process of colonisation and later socioeconomic transformations. Archaeological studies date the first inhabitants of the area to about 2,200 years ago (Ramírez and Ávila, 1994). In the Solapa Funeraria de Soroa – located in the area of the same name – food remains, and fragments of rudimentary objects used by these aborigines, as well as bones and a skull have been found. The inhabitants in this part of the island were hunters, fishermen and gatherers.

The sale of circular areas or farms throughout the country (called *hatos* and *corrales*) started in the 15th and 16th centuries (García Rodríguez, 2006); the former to raise large livestock, the latter for small livestock. These activities were located primarily within what is today the protected area. Raising small livestock was less labour intensive and so small plots were cultivated to plant subsistence crops to feed the workers in charge of the *corral*. This activity did not require any clearing of forested areas.

The French Revolution of 1789 had a direct effect on the Caribbean colonies, causing migration from the French colonies. On the one hand, the arrival of French immigrants to Cuba was stimulated by the economic transformations that were taking place on the island, and, on the other, the Spanish administration promoted white immigration, which boosted coffee plantations in the Sierra del Rosario (Ramírez and Paredes, 2000) and drove economic, social and cultural changes. However, some poor agroforestry practices were also adopted, leading to the deterioration and subsequent disappearance of coffee plantations. The authors observe that the loss of these plantations was due to the indiscriminate clearing of forested areas, "…letting themselves be carried away by the ease in knocking down the mountain and reaping only its first fruits".

In the 1820s and 1830s, the cultivation of sugarcane started to expand to the plain or hilly areas on the south-eastern outskirts of Sierra del Rosario, with the establishment of several sugar cane mills (*ingenios*) (Carambola, El Galope, Empresa and Las Delicias) in 1848 (Zamora and Zamora, 1996). These mills used firewood to evaporate the *guarapo* (liquid extracted from sugarcane) and this demand led to the widespread felling of forests in the plains and nearby mountains which – together with clearing land to plant coffee – caused serious environmental impact in what is today the protected area. The first half of this century was characterised by large estates mainly raising small livestock, but the felling of trees for firewood continued and charcoal production was widespread. Rural workers settled on scattered small plots based on subsistence farming, living in precarious conditions. These workers took advantage of the already degraded forests by clearcutting for charcoal which was the main economic sustenance of families in the mountains and eventually leading to deforested landscapes

In the context of the environmental and social degradation of this mountain area, a comprehensive development project was initiated in 1968. One objective was the reforestation of an area of 50 km², using the continuous slope terrace system, consisting in transversal cuts following the contour lines of the mountain slopes. The second objective was to build a community from the impoverished and dispersed population of the area and finally, to build road infrastructure to connect it with the rest of the country. Las Terrazas community (Figure 3.2) was inaugurated in 1971, engaging its inhabitants primarily in forestry-related activities, promoting the restoration of the forest cover and the environmental regeneration in the reforested area. The Sierra del Rosario area, including the 50 km² of reforested land, Las Terrazas community, and 16 other settlements (6,662 inhabitants) was later declared a Biosphere Reserve.

Figure 3.2 Las Terrazas community.

Current economic activities

The Biosphere Reserve and its outer area of influence host a combination of agricultural, forestry and tourism activities. While farming is an activity present throughout the entire protected area, in the montane sector to the west, small-scale subsistence farming is prevalent, with coffee plantations as the principal source of earnings alongside fruit trees and various other crops.

In the highlands (mostly dominant in the protected area) agriculture is integrated with forestry which is managed by three entities that own over 80% of all land. This activity is concentrated in the mountain range and covering areas in the municipalities of Candelaria, Artemisa and Bahía Honda. The main forest products are logs, firewood, and charcoal (Estación Ecológica, 2015), while to the north, east and south, livestock raising, with agriculture as a minor activity, predominates.

Conservation actions in the protected area combined with some developments have generated conditions to increase eco-sustainable tourism. This activity is located in two areas, in the far centre-west area of the Biosphere Reserve, near Villa Horizontes Soroa and La Caridad campsite, and in the centre near Las Terrazas complex and El Taburete campsite. Other areas such as La Chorrera campsite and Charco Azul, to the southeast, also attract visitors but less than other areas.

Interactions between human settlements and the surrounding environment

The Biosphere Reserve renews its management plan every five years. All the plans have addressed the key environmental, social and economic issues affecting local communities, by linking with community representatives who participate in the diagnosis of the protected area and plan the functional zoning. Furthermore, this is vital in elaborating action plans for the coordination, monitoring and management of resources, public access, scientific research, as well as local development. Communities, residents and related institutions all actively contribute through walking with personnel such as forest rangers to establish observation points from which to monitor changes in ecosystems and to maintain the flow of information to favour research, monitoring and protection of the protected area. This requires undertaking training of the majority of stakeholders, to allow integration in the management of the protected area. The participation of the local population is also promoted through various annual events, such as the *Festival de Aves Migratorias* (Migratory Bird Festival), *Feria Agroecológica* (Agroecological Fair), *Feria de Semillas y Agrobiodiversidad* (Seed and Agrobiodiversity Fair), and *Environmental Days* for the *World Environment Day*. The annual Seed and Agrobiodiversity Fair allow farmers to showcase and exchange their products, experiences and seeds, thus contributing to the conservation of the traditional agricultural biodiversity of the area since this exchange increases the diffusion of prime species and cultivars, as well as promoting the values of local agrobiodiversity, often recognized only at the farm level. The environmental education programme is aimed at all segments of society, although it particularly focuses on children, as the protected area has ten schools of different levels of tuition and a group of teachers.

Local and indigenous knowledge has also been an object of study with the aim of rescuing, updating and promoting its growth. This is the case of the conservation of agricultural diversity and the cultural heritage of traditional Cuban cuisine, especially when food products from the agricultural and natural diversity of the area are used as ingredients. *Mi Plato y Yo* is a culinary project that has had a major impact on Las Terrazas community – later joined by other rural communities – by reviving the cultural heritage of the mountain inhabitants (Figure 3.3).

The Board of Directors of the Biosphere Reserve, in coordination with governments and local leaders, has promoted initiatives to consolidate the tangible and intangible cultural heritage of the communities. These also include the declaration of cultural *patios*, where recovery of traditional knowledge of culinary culture and of the dances and songs of African ancestors is promoted.

The Agrobiodiversity Sub-programme, within the Management Plan of the National System of Protected Areas (CNAP, 2013), started in the Sierra del Rosario Biosphere Reserve (EESR, 2015, 2020). For the first time in Cuba, agrobiodiversity (cultivated genetic diversity on rural farms, its use, management and marketing) was included within the national plans for the conservation of natural

Figure 3.3 The culinary project Mi Plato y Yo in the Las Terrazas community.

biodiversity. This guarantees (from the scientific and planning perspectives) that agrobiodiversity may contribute to secure an adequate connectivity and exchange between the cultural and natural landscapes, and, ultimately, the conservation of the country's agricultural heritage.

However, despite these efforts, a preference by residents to produce goods and services focused on tourism has been observed. Due to its higher productivity, the communities of Las Terrazas and Soroa have redirected the farming and forestry workforce towards the tourism sector. Similarly, some rural farmers have adjusted their management of agricultural diversity, cultivating more productive species and varieties that are more commercially viable such as cassava (*Manihot esculenta*) and the common bean (*Phaseolus vulgaris*), abandoning the traditional varieties. Notwithstanding, these commercial varieties are not entirely suited to the local environmental conditions, as they are less resistant to pests and diseases, and consequently high yields are not maintained over time. A shift from agriculture to state-advocated livestock rearing (either indoor or semi-free, and especially pig farming) has also been observed on some farms driven by state agreements, thus limiting the diversification of agriculture. Based on the previous cases, the Reserve administration has proposed to strengthen training activities and the adoption of strategies for the valorisation of agrobiodiversity products that would allow farmers to find profit from the traditional biodiversity conserved on their farms.

The Sierra del Rosario Biosphere Reserve and its surrounding area are an example of the sustainable relationship between man and nature. A comprehensive development project, started in 1968, represented a watershed moment for the area. Early on, it contributed to advance the economic and social development, although at the cost of reducing the natural values. Today, without abandoning the objective of economic and social development, it places more emphasis on the conservation and recovery of ecosystem services. Thus, the Biosphere Reserve is now the result of a partnership of farmers and institutions, and a space open to transformative ideas that place the sustainable use of its resources at the centre of its policy.

Note

1 The information on climatic variables is based on data from the instruments located at the Sierra del Rosario Ecological Station. Temperature was measured in the period 1968–2004 (after this date the equipment deteriorated with only the rain gauge being maintained), precipitation was measured in the period 1968–2018, and the humidity values refers to historical data collected at the same station.

References

Capote, R.P., Menéndez, L., García, E.E., Vilamajó, D., Ricardo, N., Urbino, J. and Herrera, R. (1988) 'Flora y vegetación', in R.A. Herrera, L. Menéndez, M.E. Rodríguez, and E.E. García (eds) *Ecología de los Bosques Siempreverdes de la Sierra del Rosario. Proyecto MAB No. 1, 1974-1987*, Oficina Regional de Ciencia y Tecnología de la UNESCO para América Latina y el Caribe, Montevideo, pp110–130

CNAP (2013) 'Plan del Sistema Nacional de Áreas Protegidas 2014-2020', Ministerio de Ciencias Tecnología y Medio Ambiente, La Habana, 335pp

EESR (2015) 'Plan de Manejo Reserva de la Biosfera Sierra del Rosario 2016-2020', Estación Ecológica Sierra del Rosario, Las Terrazas, unpublished, 170pp

EESR (2020) 'Plan de Manejo Reserva de la Biosfera Sierra del Rosario 2021-2025', Estación Ecológica Sierra del Rosario, Las Terrazas, unpublished, 185pp

Furrazola, G. (1988) 'Generalidades sobre la Geología de la Sierra del Rosario', in R.A. Herrera, L. Menéndez, M.E. Rodríguez and E.E. García (eds) *Ecología de los Bosques Siempreverdes de la Sierra del Rosario, Proyecto MAB No. 1, 1974-1987*, Oficina Regional de Ciencia y Tecnología de la UNESCO para América Latina y el Caribe, Montevideo, pp96–109

García Rodríguez, M. (2006) 'Ingenios Habaneros del Siglo XVIII: Mundo Agrario Interior', *América Latina en la Historia Económica*, no 26, pp43–75

González, A.V. (1989) 'Reserva de la Biosfera Sierra del Rosario. Paisajes Físico Geográficos', in *Nuevo Atlas Nacional de Cuba*, Instituto Cubano de Geodesia y Cartografía, http://repositorio.geotech.cu/jspui/handle/1234/3471, accessed November 2021

Guerra, J.L. (2010) 'Herpetofauna de la Sierra del Rosario, Artemisa, Cuba. Checklist September 2010', Estación Ecológica Sierra del Rosario, Las Terrazas, unpublished, 6pp

Guerra, J.L. and Mancina, C.A. (2010) 'Aves de la Sierra del Rosario, Artemisa, Cuba. Checklist September 2010', Estación Ecológica Sierra del Rosario, Las Terrazas, unpublished, 4pp

Gundlach, J. (1868) 'Revista y catálogo de los mamíferos cubanos', in F. Poey (ed) *Repertorio Fisico-Natural de la Isla de Cuba*, Imprenta de la viuda de Barcina y Comp, vol II, La Habana, pp40–56

Hernández Jiménez, A. (1999) *Nueva versión de la clasificación genética de los suelos de Cuba*, Instituto de Suelos, Ministerio de la Agricultura, Cámara Cubana del Libro, La Habana, 64pp

Núñez Aguila, R. and Barro Cañamero, A. (2012) 'A list of Cuban Lepidoptera (Arthropoda: Insecta)', *Zootaxa*, vol 3384, pp1–59

Ponce de León, J.L., Rodríguez, R. and Guerra, J.L. (2010) 'Peces dulceacuícolas de la Sierra del Rosario, Artemisa, Cuba. Checklist September 2010', Estación Ecológica Sierra del Rosario, Las Terrazas, unpublished, 8pp

Ramírez, J.F. and Ávila, L. (1994) 'Historia del municipio Candelaria desde sus orígenes hasta 1986', Museo de Candelaria, Candelaria, unpublished, 250pp

Ramírez, J.F. and Paredes, F.A. (2000) *Francia en Cuba: los cafetales de la Sierra del Rosario (1790-1850)'*, Ediciones UNIÓN, La Habana, 103pp

Urquiola, A., González, L. and Novo, R. (2010) *Libro rojo de la flora vascular de Pinar del Río*, Universidad de Alicante, Alicante, 458pp

Zamora, J.L. and Zamora, M.J. (1996) 'La industria azucarera en el siglo XIX en Candelaria', Museo de Candelaria, Candelaria, unpublished, 20pp

4   The eastern mountains

Cuchillas del Toa Biosphere Reserve

Gerardo Begué-Quiala, Geovanys Rodríguez Cobas, Hayler M. Pérez Trejo and Rey F. Guarat Planche

The Cuchillas del Toa Biosphere Reserve (RBCT in the Spanish acronym) is located in the north-eastern part of Cuba and is classified as a Managed Resource Protected Area (APRM in the Spanish acronym), International Union for Conservation of Nature (IUCN) category VI, and covers an area of 208,305 ha. Located in the Nipe-Sagua-Baracoa mountain range, it is a varied and primarily mountainous landscape. The Reserve includes territories in the provinces of Guantánamo and Holguín; 74.2% of its total area in the former and 25.8% in the latter, with the remaining 6,013 ha in the marine fringe. With Decree 197/95 of 1987 it was declared a Biosphere Reserve. That same year it became part of the UNESCO International Network of Reserves of the Man and the Biosphere (MaB) Programme (Villaverde et al., 2014). The Reserve is a remnant natural area containing some of the best preserved ecosystems in the country and so its restoration and conservation is of the highest priority.

The Cuchillas del Toa PAMR includes six protected areas. The core zone of the RBCT is the category II IUCN Alejandro de Humboldt National Park (PNAH in the Spanish acronym) declared as such on December 14, 2001, covering an area of 70,680 ha, of which 2,250 ha are marine. This core zone is the most important protected area in Cuba in terms of biodiversity, conservation and ecological integrity, and hosts the largest number of species and greatest level of endemism. The RBCT also includes different areas which were classified (Begué-Quiala et al., 2022) as following: Pico Galán Managed Floristic Reserve (MFR) covering an area of 437 ha, Alto de las Canas Ecological Reserve (Puriales) (ER) covering 3,012 ha, Yunque de Baracoa Outstanding Natural Element (ONE) covering 1,921 ha, Monte Verde MFR covering 2,000 ha, and finally, Salto Fino Antonio Núñez Jiménez ONE, with an extension of 736.2 ha (Figure 4.1).

The Reserve is located in the rainiest area of the island with rainfall ranging from 1,500 to 4,500 mm/year, which is also the coolest area. Climatic conditions are determined by the interaction between prevailing winds and the local relief which causes abundant rainfall. These have led to the formation of a dense hydrographic network with 26 first-order basins, 14 of which lie entirely inside the Reserve, particularly the Toa basin one of the seven principal national basins and covering 50.9% of the RBCT.

DOI: 10.4324/9781315183886-5

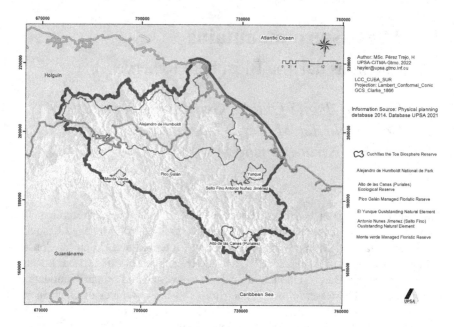

Figure 4.1 Main protected areas in Cuchillas del Toa Biosphere Reserve.

The geology – with serpentine, peridotite, karst and pseudokarst formations – is an example of the development of plant communities on ultrabasic rocks, challenging for plant survival. The protected area extension, variation in altitude, complex lithology, and diversity of landforms have generated a range of ecosystems and species unparalleled in the Insular Caribbean.

Natural values

Nineteen plant formations have been identified in the RBCT (Samek, 1974; Capote and Berazaín, 1984; Bisse, 1988; Borhidi, 1989; Reyes, 1994; Del Risco, 1995; Borhidi, 1996). These formations include low altitude rain forest, submontane rain forest, cloud forest, submontane mesophyll evergreen forest, coastal and sub-coastal microphyll evergreen forest, gallery forest, mangrove evergreen forest, *Pinus cubensis* pine forests, semi-deciduous broadleaf forests, coastal and sub-coastal xeromorphic scrub (*manigua costa*), thorny xeromorphic scrub on serpentine (*cuabal*), subspiny xeromorphic scrub on serpentine (*charrascal*), montane scrub, *mogotes* vegetation complex, rocky and sandy coastal vegetation complex, secondary forests and scrub.

The RBCT is considered one of the least explored regions of the country, as there are still areas where no collections, scientific expeditions or research have been conducted and, in some cases, the few expeditions that have ventured were primarily exploratory. One of the most studied sites is the Alejandro de Humboldt National Park, although still approximately 42% of this area has not been explored.

Despite this, more than 1,071 species from 472 genera and 123 families have been registered, although a total of 1,500 species has been estimated.

The Reserve is home to 595 species endemic to Cuba (24 of which still to be confirmed), representing around 55.5% of all reported species and 18.7% of the 3,187 vascular plants endemic to Cuba. The families with the greatest level of endemism are Rubiaceae (58 species), Euphorbiaceae (47 species), Astreaceae (45 species), Myrtaceae (40 species) and Melastomataceae (38 species). Endemism is predominant in other plant families, e.g., the Bignoniaceae and Buxaceae families (Martínez et al., 2005) with almost all species being endemic.

Interest in the protection and conservation of the natural resources of this area started in the 1960s, when two important Nature Reserves were declared: Boca de Jaguaní and Cupeyal del Norte. In the 1980s, great interest was raised by the report of *Campephilus principalis*, a Cuban endemic species previously considered extinct. This is one of the most important vertebrate species in the area and an endemic relict species highly endangered in Cuba since the other species of the same subspecies of North America and Mexico are already extinct. Subsequently, the Alto de Iberia and Ojito de Agua Fauna Refuges were declared and merged in 1996 creating what is known today as the Alejandro de Humboldt National Park.

The rivers flowing from the mountain peaks are among the fastest flowing in the Insular Caribbean. The waters have a high oxygen content which favours the presence of numerous species of freshwater fauna such as fish, crustacea and molluscs found in waters well upstream (Rodríguez-Cobas, 2015).

The presence of several anchialine species (which spend part of their life cycle in the sea and part in freshwater), such as tetids (*Sicydium* sp.), eels (*Anguilla rostrata*) and shrimps, is notable. The Reserve's fresh waters are home to more than 17 fish species, 90 freshwater shrimp species representing 47% of Cuban species, 6 species of molluscs and a still undetermined number of other invertebrates. The freshwater prosobranch mollusc *Pachychilus violaceus* is a local endemic, while the Toa river basin and its tributaries are the major habitat of the *Guaso joturo* or *biajaca* (*Nandopsis ramsdeni*), a freshwater fish, endemic and limited to the eastern part of Cuba.

Studies of the marine biota have been conducted primarily in areas of interest for management, but results are still unpublished. The continental shelf is very narrow or almost non-existent in some areas, which limits both the presence and extent of marine biotopes, as compared to other regions of Cuba, although this does not lessen their physical, ecological and socioeconomic relevance. The Reserve also encompasses the easternmost cays of the country, Cayo del Medio in the Bay of Yamanigüey and Los Cayitos in the Bay of Jaragua. Additionally, linear deltas or *tibaracones* are located at the mouths of several rivers, a unique geomorphological formation only found in river systems flowing into the northern coast of Baracoa.

To date, 1,075 marine species have been inventoried, among them molluscs (534 species), fish and sharks (152 species), cnidarians or coelenterates (88 species), crustaceans (82 species) and macroalgae (78 species). The high marine biodiversity – virtually unknown until recently – has a Caribbean biogeographic character reflecting the overspill of the marine biota of the Western Caribbean Sea into the northeastern part of the Cuban coasts. One notable feature is the presence of a small population of the West Indian manatee (*Trichechus manatus*), in the bays of Taco, Jaragua and Yamanigüey.

The inventory includes the description of 16 species of marine molluscs, previously unknown to science (Espinosa et al., 2008; Espinosa and Ortea, 2013; Espinosa and Ortea, 2014), among which *Prunum humboldti* and *Prunum tacoensis* are reported as local endemisms in Cayo del Medio, Bahía de Yamanigüey and Bahía de Taco.

Invertebrates are the most diverse and abundant group of terrestrial fauna in the Reserve. According to studies carried out in the core zone, there is an invertebrate/vertebrate population ratio of 18:1 (Villaverde et al., 2014). Terrestrial molluscs are one of the most studied and known groups of invertebrates with 45 species distributed in 16 families and 27 genera, and with 75.6% endemism. Local and sub-regional endemisms predominate, with 12 species representing 26.7% of the area's endemisms. The Camaenidae and Helminthoglyptidae families are reported with seven and six endemic species respectively, the latter including the species *Polymita picta*, one of Cuba's most unique terrestrial molluscs (Maceira, 2005).

Some 106 species of spiders have been identified, described and grouped in 32 families and 82 genera, with 13.4% endemism (Alayón and Sánchez-Ruíz, 2005), while 19 species are attributed to the Scorpiones, Amblypygi, Schizomida, Solpugida, Ricinulei and Uropygi orders (Teruel, 2005). The Diptera order has been little studied in Cuba but even so, 108 species have been identified in the area, seven of which are local endemics (Garcés, 2005).

Two hundred ninety-eight species of Hymenoptera from 35 families are also reported, representing 25.7% of the known species in the country, while about 45 species of diurnal Lepidoptera have been registered, although about 60 are estimated to be present in the area (Portuondo and Fernández, 2005). To date, 621 taxa of the above-mentioned invertebrates have been recorded.

Vertebrates, such as amphibians, reptiles, birds and mammals have been studied more extensively (Fong et al., 2005). 20 species of amphibians (90% endemic) have been recorded in the area, and 42 species of reptiles (73.8% endemic), while 142 species of birds have been recorded, 21 of which are endemic to Cuba.

16 species of mammals have been described, 5 of which are exotic species. The area has five endemisms: two bat species (*Phyllonycteris poeyi* and *Natalus primus*), two hutias, *andaraz* and *conga* (*Mesocrapomys melanurus* and *Capromys pilorides* respectively) and the Cuban solenodon or *almiquí* (*Solenodon cubanus*). Considered a living fossil, this endangered nocturnal species is rarely found and only in the best conserved habitats (Begué Quiala et al., 2005).

The Reserve area is also home to the Wilson's hawk (*Chondrohierax wilsonii*), an endemic species with such a small population that perhaps only a few pairs of individuals remain.

Human settlements and economic activities

One of the most significant activities in the reserve is nickel-cobalt and chrome mining. The RBCT is rich in nickel and other ferrous mineral deposits and has the second largest cobalt reserves in the world. Cuba exports the extracted ore making it one of the country's main socioeconomic activities. Due to the conflicting land uses, conservation strategies have been adopted and the large mining licences –La Fangosa, Las Iberias and Piloto – were declared state reserves. The consequence of this was that these mineral reserves would not be exploited, thus protecting natural resources such as water, soil and biodiversity.

Notwithstanding, mining is having a significant environmental impact on the biophysical environment, such as the loss of biota (both terrestrial flora and fauna), of the (fertile) topsoil, and the tailings and overburden of varying depths, or the exposed bedrock all together forming a "lunar" landscape (Figure 4.2).

Some of these activities are still on-going, but since the 1990s, greater emphasis has been placed on environmental protection and the conservation of natural resources. Thus, any present-day development project must be initiated with a preliminary environmental impact assessment followed by monitoring, as well as cost-benefit and feasibility analyses, to secure their sustainability.

Following environmental regulation from 2013, 41.4 ha of degraded ecosystems in the former Piloto mining licence have developed reforestation activities leading to the restoration of around 3.8% of the forests in the area (Figure 4.3).

There are 116 human settlements (communities) in the RBCT area, with about 30,673 inhabitants and a population density of 14 inhabitants/km². The largest population age group is the 15–59 age cohort, while the 0–14 age cohort is the least represented (Zabala, 2018), showing a declining birth rate and an ageing population. Communities are small, some with less than 200 residents. The largest populations are Dos Pasos and Palenque (Yateras), Yamanigüey (Moa), and Nibujón and Cayogüin (Baracoa). These rural communities have a poor communication network and roads are in poor condition (Figure 4.4).

The shortage of flat land hinders farming mechanisation so that forestry is the predominant activity. Forested land covers 46.8% of the Reserve's area. Several

Figure 4.2 (a) Lunar landscape of a mined area and (b) an area during mineral extraction.

Figure 4.3 Silvicultural work to restore a degraded forest in the Cuchillas del Toa Biosphere Reserve.

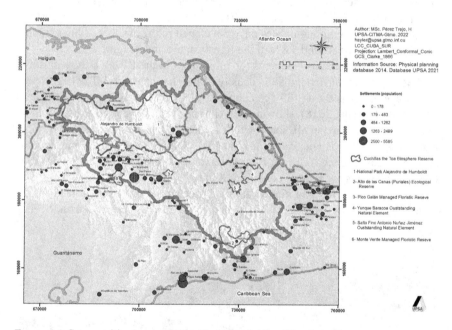

Figure 4.4 Communities located in the Cuchillas del Toa Biosphere Reserve.

cash forest formations, such as pine forest, rainforest and semi-deciduous forest have been heavily extracted, although well-preserved stands can still be found, especially in areas with poor road access in the core zone of the reserve. Farmers also cultivate coconut trees (*Cocos nucifera*), other fruit trees, coffee (*Coffea arabica*) and cocoa (*Theobroma cacao*), with high crop yields, as the basis of the local economy. Traditionally, farmers supplement cash crop income with subsistence agriculture, including seasonal crops such as maize (*Zea mays*), beans (*Phaseolus vulgaris, P. lunatus, Vigna radiata, V. unguiculata* and *Cajanus cajan*), cassava (*Manihot esculenta*), Yellow Guinea yam (*Dioscorea* sp.), sweet potato (*Ipomoea batatas*), Malanga (*Xanthosoma sagittifolium* and *Colocasia esculenta*), field pumpkin (*Cucurbita pepo*), banana (*Musa* sp.), vegetables and spices.

Coffee is an important crop in this area, with more than five commercial varieties cultivated, and Yateras is one of the municipalities in Cuba that produces the best yield of high quality coffee beans. Other crops such as oranges (*Citrus sinensis*), mandarina *(Citrus reticulata)*, sweet limes (*Citrus limetta*), breadfruit (*Artocarpus altilis*) and bananas (*Musa* sp.) are interspersed in or associated with the coffee plantations and are used primarily for family consumption.

Meat, eggs and milk are also produced, and on a smaller scale animal husbandry including large and small livestock (pigs, goats, sheep), as well as poultry. Diversification in this agrosilvopastoral system allows farmers to optimise the available resources and increase their income, especially on farms with poor access to major roads. Controlled grazing with small livestock (sheep and goats) is used on forest edges and small scattered pastures thus having very low impact on natural vegetation. Also, cultivation of various crops, such as yams, malanga, coffee, coconuts, cocoa and bananas, in the undergrowth of different forest formations has been adopted.

Shagarodsky and Castiñeiras (2013) described some vegetable species present only in this area, such as the rice bean (*Vigna umbellata*), chayote (*Sechium edule*) and American black nightshade (*Solanum americanum*), traditionally used only for selfconsumption but underutilised. Thus, the unique infraspecific variability of the RBCT has been conserved through continual use. Indeed, Fernández et al. (2007) reported unique primitive cultivars of maize such as *Morado, Pinto, Cuña* and *De Pollo*, among others. Taking into account that this is an allogamous species, farmers know how to properly proceed with seed selection to conserve the characteristics of each cultivar in successive planting cycles (Fernández Granda, 2010). The conservation of traditional crops – including both species and their infraspecific variabilities – has become a key objective in the management of the Reserve.

Farmers living in the Reserve use sustainable practices such as applying organic fertiliser produced on their farms, and crop rotation to reduce the spread of pests and diseases. Even so, soil erosion or degradation has been observed, partly associated to farming in rugged and steep terrain, but also to deforestation and other unsustainable agricultural practices.

Protected areas account for only 10% of the surface area of the planet (Primack et al., 2001), and consequently, are not sufficient to conserve global biodiversity. To address this challenge, the conservation and protection of tracts with remnants of native vegetation in cultivated areas should be promoted, as well as ecological restoration to connect these areas.

This chapter summarises the natural values of the Cuchillas del Toa Biosphere Reserve, the main threats to its conservation and how rural communities in the area integrate agricultural production with conservation. Strengthening environmental education programmes and connecting farmers and markets, particularly to sell traditional agrobiodiversity crops conserved and grown on their farms, are fundamental strategies for promoting conservation and the sustainable use of resources in a Biosphere Reserve.

Acknowledgment

The authors wish to thank Bárbaro Zabala Lahítte, Rolando Villaverde López, Oscar Caraballo Elías, Aysel García de la Cruz, Norvis Hernández, Porfilio Correa López, Elsa M. Vignón Lazo and Roermis Ortíz Argüelles, for their contribution to data collection and the writing of this chapter.

References

Alayón, G. and Sánchez-Ruíz, A. (2005) 'Las arañas', in A. Fong, D. Maceira, W.S. Alverson and T. Wachter (eds), *Cuba: Parque Nacional Alejandro de Humboldt, Rapid Biological Inventories Report no 14*, The Field Museum, Chicago, pp84–87

Begué Quiala, G. and Delgado Labañino, J.L. (2005) 'Mamíferos', in A. Fong, D. Maceira, W.S. Alverson and T. Wachter (eds), *Cuba: Parque Nacional Alejandro de Humboldt, Rapid Biological Inventories Report no 14*, The Field Museum, Chicago, pp109–112

Begué-Quiala, G., González Rivera, D., Zabala Lahitte, B., Pérez Trejo, H.M., Guarat Planche, R.F. and Imbert Planas, J.R. (2022) 'Plan Operativo Especial. Sistema Nacional de Áreas Protegidas', CITMA, Guantánamo, unpublished, 37pp

Bisse J. (1988) 'Árboles de Cuba', Editorial Científico-Técnica, La Habana

Borhidi, A. (1989) 'The main vegetation units of Cuba', *Acta Botanica Hungarica*, vol 33, no 3, pp151–185

Borhidi, A. (1996) 'Phytogeography and vegetation ecology of Cuba', Akadémiai Kiadó, Budapest

Capote, R. and Berazaín, R. (1984) 'Clasificación de las formaciones vegetales de Cuba', *Revista del Jardín Botánico Nacional*, vol 5, no 2, pp27–75

Del Risco, E. (1995) 'Los bosques de Cuba: Su historia y características', Editorial Científico Técnica, La Habana

Espinosa, J. and Ortea, J. (2013) 'Nuevas especies de la familia Marginellidae (mollusca: gastropoda: prosobranchia) de cuatro islas del Caribe: Cuba, Curazao, Guadalupe y Martinica', *Revista Academia Canaria de Ciencias*, vol XXV, pp195–218

Espinosa, J. and Ortea, J. (2014) 'Nuevas especies de moluscos Gaterópodos (mollusca: gastropoda) del Parque Nacional Alejandro de Humboldt, Sector Baracoa, Guantánamo, Cuba', *Revista Academia Canaria de Ciencias*, vol XXVI, pp107–135

Espinosa, J., Ortea, J. and Moro, L. (2008) 'Nueva especie de marginela del género Prunnum Herrmannsen, 1852 (Mollusca: Neogastropoda: Marginellidae) del Parque Nacional Alejandro de Humboldt, Sector Baracoa, Cuba', *Revista Academia Canaria de Ciencias*, vol XX, no 4, pp19–22

Fernández, L., Castiñeiras, L., Fundora, Z., Barrios, Shagarodsky, T., Cristobal, R., Barrios, O., Fuentes, V., Moreno, V., Fuentes, V., Moreno, V., León, N., García, M., Giraudi, Félix, M., Guevara, C., Acuña, G. and Puldón, G. (2007) 'Manejo dinámico de maíces tradicionales en fincas de dos áreas rurales de Cuba', *Agrotecnia de Cuba*, vol 31, no 2, pp321–327

Fernández Granda, L. (2010) 'Identificación de razas de maíz (Zea mays L.) presentes en el germoplasma cubano', Editorial Universitaria, La Habana

Fong, G.A., Díaz, L.M. and Viña, N. (2005). 'Anfibios y reptiles', in A. Fong, Maceira, D., W.S. Alverson and Wachter, T. (eds) *Cuba: Parque Nacional Alejandro de Humboldt, Rapid Biological Inventories Report no 14*, The Field Museum, Chicago, pp93–101

Garcés, G.G. (2005) 'Los dípteros', in A. Fong, D. Maceira, W.S. Alverson and T. Wachter (eds) *Cuba: Parque Nacional Alejandro de Humboldt, Rapid Biological Inventories, Report no 14*, The Field Museum, Chicago, pp89–91

Maceira, F.D. (2005) 'Moluscos terrestres', in A. Fong, D. Maceira, W.S. Alverson and T. Wachter (eds) *Cuba: Parque Nacional Alejandro de Humboldt, Rapid Biological Inventories Report no 14*, The Field Museum, Chicago, pp81–84

Martínez, E., Fagilde, M.C., Alverson, W., Vriesendorp, C. and Foster R.B. (2005) 'Plantas espermatofitas', in A. Fong, D. Maceira, W.S. Alverson and T. Wachter (eds) *Cuba: Parque Nacional Alejandro de Humboldt, Rapid Biological Inventaries Report no 14*, The Field Museum, Chicago, pp79–81

Portuondo, F.E. and Fernández, J.L. (2005), 'Los himenópteros', in A. Fong, D. Maceira, W.S. Alverson and T. Wachter (eds), *Cuba: Parque Nacional Alejandro de Humboldt, Rapid Biological Inventories Report no 14*, The Field Museum, Chicago, pp91–93

Primack, R., Rozzi, R., Feinsinger, P. (2021) 'Establecimiento de áreas protegidas', in R. Primark, R. Rozzi, P. Feinsinger, R. Dirzo and F. Massardo (eds), *Fundamentos de conservación biológica: Perspectivas latinoamericanas*, Fondo de Cultura Económica, México D.F., pp449–475

Reyes, O.J. (1994) 'Algunas consideraciones sobre la biodiversidad cubana, en énfasis con la flora fanerógama', in Sánchez, B. (ed) *Memorias del ciclo de conferencias México-Cuba*, Instituto Politécnico Nacional de México, México, D.F., pp102–129

Rodríguez-Cobas, G. (2015) 'Bases para un manejo integrado de la Zona Marino-Costera del Parque Nacional Alejandro de Humboldt, Sector Baracoa, Cuba', Master thesis, Centro de Estudios Multidisciplinarios de Zonas Costeras, Universidad de Oriente, Santiago de Cuba

Samek, V. (1974) 'Elementos de silvicultura de los bosques latifolios', Instituto Cubano del Libro, La Habana

Shagarodsky, T. and Castiñeiras, L. (2013) 'Especies de plantas subutilizadas en Cuba', *Agrotecnia de Cuba*, vol 37, no 1, pp18–25

Teruel, R. (2005) 'Otros arácnidos', in A. Fong, D. Maceira, W.S. Alverson and T. Wachter (eds), *Cuba: Parque Nacional Alejandro de Humboldt, Rapid Biological Inventories Report no 14*, The Field Museum, Chicago, pp87–89

Villaverde, R., Begué, G., B. Giraudy, C., Pérez, H.M., Ubals, R., Acebal, R., Joubert, Y., Hernández, N., Correa, P., Sánchez, L.N., Medina, A., Guarat, R.F., Balón, C.Y., Maury, O., Rodríguez, G, Matos, R., Delgado, J.L., Imbert, J.R. and López, J.B. (2014) 'Plan de manejo quinquenio 2014-2020', Parque Nacional Alejandro de Humboldt, Unidad Presupuestada de Servicios Ambientales Alejandro de Humboldt (UPSA), CITMA, Guantánamo, 113pp

Zabala, B. (2018) 'Ordenamiento Ambiental de la Reserva de Biosfera Cuchillas del Toa', Doctoral thesis, Facultad de Geografía, Universidad de la Habana

5 The lagoons

Peninsula de Guanahacabibes Biosphere Reserve

Lázaro Márquez Llauger, José Alberto Camejo Lamas, Osmani Borrego Fernández and Lázaro Márquez Govea

Introduction

The Guanahacabibes Peninsula Biosphere Reserve (RBPG) covers an area of 156,202 ha in the western part of the Cuban archipelago with coasts in the Caribbean Sea and the Gulf of Mexico. Its distinctive location determines its singular nature which is characterized by unique species, habitats, ecosystems and landscapes (Figure 5.1).

Figure 5.2 shows the functional zoning of the RBPG. The core zone covers 52,126 ha, of which 27,681 ha are marine areas. The Reserve includes the Guanahacabibes National Park, the Ciénaga de Lugones Fauna Refuge, and the Banco de San Antonio Outstanding Natural Feature. The buffer zone covers 97,941 ha (of which 26,577 ha are marine) and a terrestrial transition zone of 6,135 ha (Márquez et al., 2021).

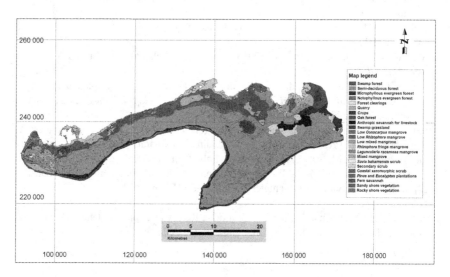

Figure 5.1 Vegetation and land cover of the Guanahacabibes Peninsula Biosphere Reserve.
Source: Marquez et al. (2021).

DOI: 10.4324/9781315183886-6

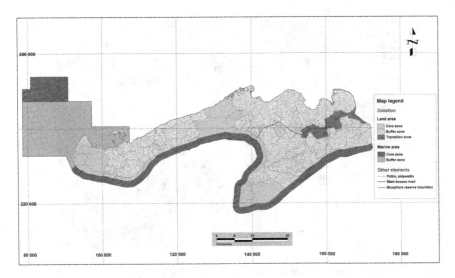

Figure 5.2 Functional zoning of the Guanahacabibes Peninsula Biosphere Reserve.
Source: Marquez et al. (2021).

Ecosystems and natural biodiversity in the RBPG

The Reserve has a variety of ecosystems and habitats, including semi-deciduous notophyll forest, evergreen microphyll forest, evergreen notophyll forest, swamp forest, mangrove forest, coastal and subcoastal xeromorphic scrub, sandy coastal vegetation complex, rocky coastal vegetation complex, and swamp grassland. Figure 5.3 shows the sandy coastal vegetation of a locality in the RBPG (Bajada de Valentín).

With a complex biogeography, the RBPG has a high floristic diversity, a fair representation of the main habitats and a good conservation status achieved through a development model favouring diverse uses of natural resources and sustainable management practices. Taking into account the number of endemic taxa, the terrestrial flora forms a distinct phytogeographic district that contains 11.1% of the endemic species reported in the country (Berazaín et al., 2005).

The floristic inventory lists 130 families, 456 genera and 791 species. The most represented families are Fabaceae and Poaceae (with 62 species each), Orchidaceae (39), Rubiaceae (35) and Asteraceae (33), representing 29.2% of all species. The most notable genera for the number of species, are *Erythroxylum, Ficus, Paspalum* and *Tillandsia* (9 species each), *Eugenia, Malphigia* and *Tabebuia* (8 species each) and *Casearia* (7 species), with an endemism of 20% (15 species), including *Harrisia taetra, Amaranthus minimus* and *Broughtonia cubensis* (Ricardo et al., 2016).

There is a high number of plants used by local communities with a broad range of uses. An ethnobotanical study conducted by Rosete et al. (2013) identified 390 medicinal species, 243 used for timber, 165 ornamental, 162 melliferous and 120 species for magic and medicinal purposes, 108 edible for humans and

Figure 5.3 Sandy coastal vegetation and fossil bluff at Bajada de Valentín.

104 species edible for animals, 40 species used as firewood, 26 used for dyeing and 18 used to obtain resin, many of them with multiple uses.

Studies of the Guanahacabibes fauna increased in the mid-1960s. From this point the first fauna inventories were published, and the current list includes 15 species of amphibians found in the Peninsula (Rodríguez-Schettino and Rivalta, 2003; Díaz and Cádiz, 2008), 35 reptiles, including endemic species of great eco-logical importance such as the Cuban iguana (*Cyclura nubila*) and the *majá de Santa María* (*Epicrates angulifer*) (Rodríguez-Schettino and Rivalta, 2003).

In the birds class 213 species have been identified, including Cuban endemisms such as the *zunzuncito* (*Calypte helenae*), *cartacuba* (*Todus multicolor*), blue-headed quail-dove (*Starnoenas cyanocephala*) and the Cuban parrot (*Amazona leucocephala*), among others (Pérez et al., 2011). The mammal community com-prises 24 species, of which 14 bats, including the threatened endemic *Natalus pri-mus* (Borroto-Páez and Mancina, 2011).

In the marine zone, 203 species of reef fish have been reported so far (Cobián et al., 2011), 42 species of corals (Perera et al., 2013), 27 gorgonians and 39 sponges (Márquez et al., 2021). The area hosts populations of endangered species such as the *cobo* (*Lobatus gigas*), the *cigua* (*Cittarium pica*) and other species with high economic value such as the spiny lobster (*Panulirus argus*).

Studies conducted in the last decade have updated the list of marine molluscs in the area to 1,016 identified species, which is the most extensive and complete inventory of marine malacofauna in a protected area in Cuba to date. This wealth of molluscs – the vast majority with planktotrophic larval development – highlights the importance of Guanahacabibes as a centre of larval dispersal of Antillean and Caribbean species into the Gulf of Mexico, cays and continental coasts of North America, attributing an importance to the Reserve that extends well beyond its national borders (Espinosa et al., 2012).

The coastal zone of the RBPG has a special significance for the conservation of threatened marine reptiles such as the *caguama* (*Caretta caretta*), the green turtle (*Chelonia mydas*) and the *carey* (*Eretmochelys imbricata*) which use the local sandy beaches to deposit their eggs (Azanza et al., 2014).

Given their degree of representation, uniqueness, level of endangerment and overall conservation value, several ecosystems have been identified where further protection and management is required. Protected forest covers a large continuous area in the core of the Peninsula and include a diverse mix of species, among which *Cedrela odorata, Drypetes alba, Cordia gerascanthus, Oxandra lanceolata, Callophyllum antillanun, Coccoloba diversifolia, Gymnanthes lucida, Bursera simaruba, Caesalpinia violacea, Calophyllum antillanun, Mastichodendron foetidissimum*. These species are in different strata, although occasionally there is some evidence of damage caused by the frequent impact of hurricanes, especially at the edges of forest stands.

Mangrove forests cover large areas in the northern part of the Peninsula with species such as red mangrove (*Rhizophora mangle*), stunted mangrove (*Conocarpus erectus*), and mixed mangroves with *Avicennia germinans* and *Laguncularia racemosa* as the dominant species. These forests are strips of variable width ranging from 20 m to 7 km, primarily along the north coast.

The sandy shore vegetation complex stretches along the southern coast of the Peninsula as a 10–100 m wide strip over a carbonate sandy substrate. This complex is dominated by *Coccoloba uvifera, Thrinax radiata* and *Bursera simaruba*. The shrubs do not exceed 2 m in height and *Suriana maritima* and *Tournefortia gnaphalodes* are also common. The area is very prone to hurricanes that can cause massive uprooting of trees, damage to branches and canopies and vegetation mortality.

On the north coast, seagrass formations are primarily composed of the phanerogams *Thalassia testudinum, Syringodium filiforme, Halodule wrightii* and *Halophila engelmannii*. They are attached to the mangrove roots on various types of sea bottoms. The common substrates are sandy, muddy, sandy-muddy and rocky with sand and mud deposits. Many marine species use this ecosystem to complete their life cycles, and they are also a favoured site for feeding and refuge for certain endangered species, such as the manatee (*Trichechus manatus manatus*) and sea turtles *Chelonia mydas* (green turtle), *Caretta caretta* (loggerhead) and *Eretmochelys imbricata* (hawksbill sea turtle).

Coral reefs extend mainly along the southern coast of the Peninsula. Existing regulations ensure that these ecosystems remain healthy and host a rich variety of species. The karst of the Peninsula also contributes by minimizing any impact associated with river water discharges and sedimentation from terrestrial sources. Forty-two species of stony corals have been recorded to date. The percentage of live coral substrate cover is high, reaching between 15% and 30%. Complex formations of *Orbicella annularis, O. faveolata, O. franksi* and *O. cavernosa* are recognized as the primary components of the deep reefs. The presence of these species is an indicator of the good overall health and favourable natural conditions for the survival of the reef. There is very little evidence of disease and low mortality values are being registered. These areas are considered among the most diverse and best-conserved not only in Cuba but in the entire Caribbean region (Perera et al., 2013). A distinctive feature of these reefs is the rich variety of fish on its slopes when compared to other Caribbean regions (Perera et al., 2013), with 201 species registered. The protection provided ensures healthy populations of commercially important species, including the *cherna* (*Epinephelus striatus*) and the *pargo criollo* (*Lutjanus analis*), as well as diverse species of conservation interest such as the parrotfish (*Sparisoma* spp.; *Scatus* spp.) and *barbero* (*Acanthurus* spp.), which play a fundamental role in food chains, algal growth control and the health of these ecosystems.

Local communities and the natural environment

There are 1,684 inhabitants living in five rural communities located in the eastern region of the RBPG where most agricultural activity is located due to the availability of arable land and water for irrigation (ONEI, 2019).

Most of the local population resides in the settlements of Malpotón and La Jarreta, where they grow dark and blond tobacco (*Nicotiana tabacum*), *viandas*, vegetables, grains and fruit, as well as raising livestock. The inhabitants of the El Valle, El Vallecito and La Lima communities are mainly involved in forestry-related activities, fruit growing and beekeeping.

Local communities play a significant role in the management of the Reserve which is channelled through the Community Environmental Groups in each community with formal and informal leaders and the support of the local government. This has been reflected in planning which has benefitted from the active participation of local community representatives in workshops and meetings to perform diagnosis, and functional zoning in the Reserve (Figure 5.2) and to design action plans for its management. Women, young people and other residents representing the different groups in the community, with better understanding of the natural ecosystem, are involved in the planning activities for the protection and conservation of the area.

The implementation of an Environmental Education Programme has reinforced the participatory approach through various initiatives and annual events, including the Community Festival for the Conservation of Sea Turtles, the Migratory Bird Festival, the Agroecological Fair, the Fishing Tournament of *Pez león* (*Pterois antennata*) and the World Environment Day. These events are held annually in different communities of the Reserve and performances by children and young people from the local communities play a key role.

An important contribution by local communities is done through the implementation of monitoring and conservation programmes. The past twenty years have seen the systematic and growing participation of local people as volunteers in the monitoring and conservation of sea turtles, the study and management of ocean debris, and the control of invasive exotic species. These activities have mainly involved young people under the guidance of local leaders, with the advice of the Reserve's experts.

The functional zoning of the RBPG and the regulation of land use and management of natural resources include the elaboration of guidelines for ecosystem protection to favour biological connectivity and sustainable development. One example is the transition zone concentrated in the central-eastern region of the RBPG, where sustainable forms of resource exploitation are encouraged and practiced. This area is home to local communities and the region's agricultural development programmes, mainly focused on the production of food. Thus, zoning recognizes the role of non-intensive farming land and agricultural biodiversity acting as biological corridors in the Reserve, as they connect relevant conservation sites in the buffer zone.

Local economy in transformation

In the last decade, important changes have taken place in the local economy in line with transformations taken place in the country's development model. Beekeeping has increased through the expansion of areas dedicated to farming and more efficient management, which have led to a considerable increase in the production of honey, wax, pollen and other by-products. Since 2017, an EU-based certification entity has issued organic certification for Guanahacabibes honey and also for its associated production system, thus attributing added value to this traditional practice which is reflected in the increase in sale prices and income for the villagers.

In the agricultural sector, the major changes have been observed in the transformation of the region's productive matrix, with more than 60 ha of land previously dedicated to the traditional cultivation of dark and blond tobacco now earmarked to increase the cultivation of *viandas*, such as *yuca* (*Manihot esculenta*) and *boniato* (*Ipomoea batatas*), vegetables including pumpkin (*Cucurbita* sp.), grains such as rice (*Oryza sativa*), beans (*Phaseolus* spp. and *Vigna* spp.) and maize (*Zea mays*), as well as fruit trees including guava (*Psidium guajava*) and *fruta bomba* (*Carica papaya*). Over the last decade, training has been provided to local farmers to favour the increase their use of organic fertilizers and biological pest control. Advances have been made in the introduction of new and more efficient irrigation technologies and the sustainable management of land in demonstration areas.

In the forestry sector, the most important change regards the reduced volume of timber extracted from natural forests, with a decrease from 18,785 m^3 in 2012 to 11,460 m^3 in 2021. In the same period, reforestation areas expanded, and agroforestry farms were promoted. In the previous decade, an area dedicated to cattle farming in the transition zone was converted into secondary forest through ecological restoration. The effect of this was a significant increase in nature tourism, mainly in diving, water sports, wildlife watching and hiking. From 12,971 in 2010, the number of tourists increased to 23,834 in 2019. Private tourism initiatives offering accommodation have been started in the La Bajada community. These new services have significantly diversified the local economy by creating an offer for the tourism sector and generating new jobs for the local population.

Agrobiodiversity

The Agrobiodiversity Programme included in the RBPG Management Plan (Márquez et al., 2021) recognizes the genetic agrobiodiversity, which is adapted to the poor soils of the area (Figure 5.4), grown and passed down from one generation to the next for both home consumption and for sale. From the ecological and planning perspectives, this ensures adequate linkage between the natural and cultural landscapes and thus, favours the conservation of agricultural heritage and significantly contributes to the nation's food sovereignty.

Vegas are large, flat and fertile areas located in the central-eastern region of the Reserve. These are managed by farmers using traditional tillage techniques grouped in a Cooperativa de Créditos y Servicios (CCS) and in a Cooperativa de Producción Agropecuaria (CPA). The main crop is blond tobacco for sale to the state and, to a lesser extent, dark tobacco for local consumption. In the *vegas*, farmers alternate tobacco harvesting with the cultivation of other crops such as pumpkin or root crops (*Ipomoea batatas, Manihot esculenta, Colocasia esculenta* and *Xanthosoma sagittifolium*). Tomatoes (*Solanum lycopersicum*) are harvested for the traditional production of tomato puree, which is preserved from one year to the next. Grain crops include the *frijol común* (*Phaseolus vulgaris*), sesame (*Sesamum indicum*) and maize (*Zea mays*).

In the transition area of the RBPG there are small agroforestry farms located in the central part of the Peninsula, where fruit species such as mango (*Mangifera indica*), guava and papaya are cultivated, as well as different citrus fruits such as key lime (*Citrus aurantifolia*), sweet orange (*Citrus sinensis*) and mandarine (*Citrus reticulata*), and to a lesser extent other more customary crops such as coffee (*Coffea arabica* and *C. canephora var. robusta*). Farm production is oriented towards domestic consumption, the provision of animal feed and direct marketing to the public through local markets. In recent years, the direct sale of food and fruits to local tourist organizations has also developed.

Figure 5.4 Cultivation of papaya (*Carica papaya*) on land with poor soils, showing rocky outcrops, in the La Lima community.

Figure 5.5 Fruit trees in an allotment in the El Valle community.

Home gardens are the basic pillar for domestic consumption (Figure 5.5). They are located next to the farmers' houses and – depending on the type of crop – often have live fences to protect crops from domestic animals. They provide families with fresh, short-cycle vegetables, *viandas* and condiments, such as varieties of *ají cachucha* (*Capsicum chinense*), *ají guaguao* (*Capsicum frutescens*), pepper (*Capsicum annuum*), garlic (*Allium sativum*) and onion (*Allium cepa*). Medicinal plants such as the *caña santa* (*Costus pictus*), *flor de España* (*Lippia alba*) and carpenter bush (*Justicia pectoralis*), as well as a wide variety of fruit trees such as mango, guava (*Psidium guajava*), pineapple (*Ananas comosus*) and cherimoya (*Annona cherimola*), among others. Figure 5.5 shows some fruit trees in an allotment in the El Valle community.

One interesting practice by communities in the transition zone are the so-called *tumbas*. These are small family plots located outside communities generally infested by invasive alien species, which farmers remove mechanically, while preserving natural vegetation. After soil tillage, cassava, sweet potato, malanga, papaya, plantain (*Musa* sp.), guava and other *viandas*, vegetables and fruits are grown for family consumption. In recent years, *tumbas* have also been extended into semi-flooded areas for rice cultivation and have become an important source of food that has improved the food security of farming families.

As part of the diagnosis in the RBPG Management Plan (Márquez et al., 2021) traditional knowledge of agricultural biodiversity and management practices applied in soils of low fertility with minimum tillage methods has been compiled and synthesized. An inventory was conducted in farmland (*vegas*, agroforestry farms, *tumbas* and home gardens) through surveys with farmers and residents as target population. As a result of these 213 species cultivated by local inhabitants were identified, illustrating the high agricultural biodiversity, supplemented by a further 26 wild species used by the communities.

An agroecological fair is held annually in the community of La Jarreta to disseminate and exchange traditional knowledge. In this event, which coincides with Earth Day, local farmers showcase relevant experiences, display and exchange agricultural products, seeds and typical rural Cuban dishes. In this way, agroecological fairs contribute to empower local farmers and have an impact on the family economy and knowledgebase.

The adoption of a participatory method in the planning of the RBPG has been the cornerstone to encourage the inclusion of local communities in programmes for the conservation and management of agricultural biodiversity. In this sense, the increasing role played by farmers and their families in selecting actions to be adopted by the Management Plan has helped strengthen their commitment to these objectives as well as their sense of belonging to the territory, which altogether involves a significant social impact.

The identification of key targets to promote the conservation and sustainable use of agricultural biodiversity in the 2022–2026 Management Plan was done through these participatory actions (Márquez et al., 2021). Fundamentally, the objectives were to update the inventory of both cultivated and wild species, to re-examine the ecosystem services provided by agricultural biodiversity, including the role of farmland as biological corridor, and to promote the commercialization of traditional agricultural biodiversity products in local markets and tourist facilities. It additionally seeks to reduce unsustainable agricultural practices which hamper the conservation of natural biodiversity, and to empower local decision-makers in the adoption of sustainable development strategies that pursue to strengthen the combined conservation of agricultural biodiversity and natural biodiversity.

References

Azanza, J., Ibarra, M.E., Espinosa, G., Cobián, D., Angulo, J., Forneiro, Y., González, G., Márquez, L., Campohermoso, E. and Hernández, N. (2014) 'Estudio y conservación de las tortugas marinas que anidan en la Reserva de la Biosfera Península de Guanahacabibes', *Revista Anales de la Academia de Ciencias de Cuba*, vol 4, no 2, pp1–18

Berazaín, R., Areces-Berazaín, F., Lazcano, J.C. and González, L.R. (2005) 'Lista roja de la flora vascular cubana', *Documentos del Jardín Botánico Atlántico*, vol 4, no 1, pp1–86

Borroto-Páez, R. and Mancina, C.A. (2011) *Mamíferos en Cuba*, UPC Print, Vaasa, Finlandia, 271pp

Cobián, D., Claro, R., Chevalier Monteagudo, P., Perera, S. and Caballero, H. (2011) 'Estructura de las asociaciones de peces en los arrecifes coralinos del APRM Península de Guanahacabibes, Cuba', *Revista Ciencias Marinas y Costeras*, vol 3, no 1, pp153–169

Díaz, L.M. and Cádiz, A. (2008) 'Guía taxonómica de los anfibios de Cuba', *Abc Taxa*, vol 4, pp1–294

Espinosa, J., Ortea, J., Sánchez, R. and Gutiérrez, J. (2012) 'Moluscos marinos de la Reserva de la Biosfera de la Península de Guanahacabibes', Instituto de Oceanología, La Habana, 325pp

Márquez, L., Borrego, O., Cobián, D., Cardoso, C., Sánchez, J.I., Camejo, J.A., Linares, J.L., Varela R. and Sosa, A. (2021) 'Plan de Manejo de la Reserva de la Biosfera Península Guanahacabibes para el periodo del 2022 al 2026', CITMA-ODIG, La Habana, 227pp

ONEI (2019) 'Anuario Estadístico de Sandino', Oficina Nacional de Estadísticas e Información, La Habana, 87pp

Perera, S., Alcolado, P., Caballero, H., De la Guardia, E. and Cobián, D. (2013) 'Condición de los arrecifes coralinos del Parque Nacional Guanahacabibes, Cuba', *Revista Ciencias Marinas y Costeras*, vol 5, no 1, pp69–86

Pérez, A., González, H., Belliure, J. and Delgado, F. (2011) 'Comunidades de aves en la Reserva de la Biosfera Península de Guanahacabibes, Cuba: Dinámica recuperativa después de aprovechamiento forestal', *Journal of Caribbean Ornithology*, vol 24, no 1, pp26–31

Ricardo, N., Herrera, P.P., Echevarría, R., Rosete, S., Hernández, A. and Álvarez, A.D. (2016) 'Península de Guanahacabibes, Pinar del Río, Cuba I. Flora', *Acta Botánica Cubana*, vol 215, no 1, pp114–140

Rodríguez-Schettino, L. and Rivalta, V. (2003) 'Lista de especies', en L. Rodríguez-Schettino (ed) *Anfibios y Reptiles de Cuba*, UPC Print, Vaasa, Finlandia, pp162–165

Rosete, S., Ricardo, N., Escarré, A., Herrera, P., Aguilar, A., Vergara, L., Medina, R., Hernández, J.A., Castañeira M.A. and Vega, S. (2013) 'Recursos vegetales en la Reserva de la Biosfera Península de Guanahacabibes, Cuba', *Revista Cubana de Ciencias Forestales*, vol 1, no 2, pp121–132

6 On the edge of Sierra Maestra

Baconao Biosphere Reserve

Giraldo Acosta Alcolea

The Biosphere Reserve

This Protected Area of Managed Resources (PAMR) in Cuba's National System of Protected Areas (CNAP, 2013) was declared by UNESCO as the Baconao Biosphere Reserve (BBR) in 1986 and designated as a World Heritage Site in 2001. It has an extension of 84,887 ha of terrestrial land and maritime area, stretching over more than 66% of the municipality of Santiago de Cuba. With 38,000 inhabitants living in 52 settlements, this is the Reserve with the highest population density in the country.

The Reserve is located in the eastern part of the Sierra Maestra, and extends through the Sierra de la Gran Piedra, Santa María del Loreto and part of the coastal terraces of the southern Sierra Maestra, in the Mar Verde-Baconao sector (Figure 6.1). It corresponds to the El Cobre group and the Jaimanitas, Maya, La Cruz, Caney, Hongolosongo, San Luis and Puerto Boniato formations, consisting of tuffs, andesitic lavas, massive organodetritic limestone, compact coralline beds, aleurolites and sandstones.

Geomorphologically, weak erosional-denudation processes that generate gravitational (landslides) and fluvial denuded formations were reported by Salmerón and Álvarez (2014). The climate is predominantly cool, with an average annual temperature between 18°C and 24°C, while the annual relative humidity varies on average between 75% and 95%. Red leached ferrallitic and yellowish ferrallitic soils are predominant.

The ecological value of its flora and fauna, as well as its high level of endemism, is relevant. The flora includes 2,000 species of phanerogams (14.0% endemic and 44 threatened species), 496 ferns (5.4% endemic) and 277 bryophytes (1.4% endemic). It is important to note that more than 50% of the reported phanerogam flora has medicinal, timber, folkloric or edible uses (Salmerón and Álvarez, 2014).

There are 18 vegetation formations in the area, of which the most important (due to their extension and ecological relevance) are montane rainforest, pine forest, mesophyll evergreen forest, mesophyll semi-deciduous, microphyll semi-deciduous and secondary forest, as well as coastal and pre-coastal scrub and secondary vegetation.

The fauna consists of 140 species of arachnids of the Araneae order with 19% endemism, 120 species of butterflies (63% of the species present in Cuba and 64% endemism) including some rare species, such as *Anaea cubana*, *Astraptes habana*

DOI: 10.4324/9781315183886-7

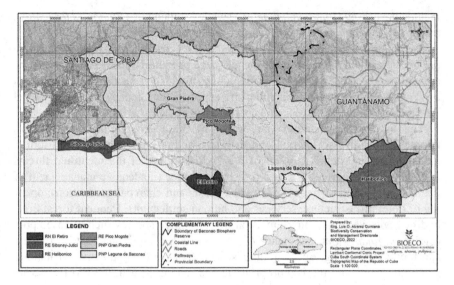

Figure 6.1 Location of the Baconao Biosphere Reserve.

Source: Álvarez (2022).

and *Parides gundlachianus*. Thirty-eight species of terrestrial molluscs have been catalogued from 15 families and 27 genera, with 81% endemism, 8 of which endemic to the eastern region, while one species classified as critically endangered.

In addition, 94 fish species identified in the area belong to 38 families and include two vulnerable species: the queen triggerfish and the gag grouper (*Balistes vetula* and *Mycteroperca microlepis* respectively).

The amphibian group includes 15 endemic species (of the 17 species existing in Cuba) from four families of the Anura order, the only one in the country. This group has a high level of endemism (88%), with one local endemic species. Nine of the Reserve's endemic species are listed on the IUCN Red List (2019) in the Endangered and Vulnerable categories.

Of the 42 reptile species identified, 14 are exclusive to the eastern region, with a 45% degree of regional endemism. Three species with a restricted distribution are noted: *Sphaerodactylus ramsdeni*, *S. schwartzi* and *S. siboney*. This group has a high level of endemism (74%) in the BBR, which accounts for 28% of all Cuban terrestrial reptiles. About 24% of these species are endangered and listed in the *Red Book of the Vertebrates of Cuba* (González et al., 2012), while five have also been included in the IUCN Red List (2019) under the Endangered and Vulnerable categories: *Anolis rejectus* (Santiago grass anole), *Cyclura nubila* (Cuban iguana), *Tropidophis pilsbryi* (Cuban white-necked dwarf boa), and *Sphaerodactylus dimorphicus* and *S. torrei* (Cuban broad-banded geckolet).

A high level of endemism is also noted for the avifauna. Of the 153 species, 13 are endemic (out of 22 in Cuba), representing 59% endemism. The area is part of an important migratory route for North American birds of prey which include the osprey (*Pandion haliaetus*), swallow-tailed kite (*Elanoides forficatus*), merlin (*Falco columbarius*), and peregrine falcon (*Falco peregrinus*).

The mammals community comprises 24 species, of which 17 are native (14 chiroptera and 3 rodent species) and 7 have been introduced. The chiropterans are well represented, with 13 species – some of which are endemic to Cuba such as *Phyllonycteris poeyi* – inhabiting the caves of the Siboney-Juticí Ecological Reserve, one of the Reserve's core conservation nuclei.

The integrity, composition and value of these ecosystems, however, are threatened by diverse internal and external factors influencing the utilisation, management and administration of the PAMR. Some of the identified threats include: mining, agricultural and forestry activities (inadequate management of plantations, felling of trees for firewood, posts and charcoal production), opening of hotel infrastructure, services and human settlements, frequent forest fires, hydrocarbon waste dumping, invasion of non-native species such as rose apple (*Syzygium jambos*), aroma blanca (*Leucaena leucocephala*) and catfish (*Claria* spp.), coral bleaching, overfishing, destruction of corals by underwater fishing, anchoring boats and the inappropriate use of fishing gear in the PAMR marine areas.

Agricultural activity, conservation and management of agrobiodiversity and natural resources

Agriculture is one of the main economic activities in the Reserve and its buffer zone. Several state-owned companies play an important role as they own more than 85% of the total area of the PAMR. The Gran Piedra-Baconao Agroforestry Company is the main landowner, with numerous farms and forestry estates. Likewise, the Empresa Pecuaria Caney is engaged in cattle raising and the Empresa Nacional para la Protección de la Flora y la Fauna is involved – among other activities – in raising poultry and pigs in farming areas. Different production systems, such as the Cooperativas de Producción Agropecuaria (CPA), Cooperativas de Créditos y Servicios (CCS), and the Unidades Básicas de Producción Cooperativa (UBPC), can be found. Additionally, usufructuaries and small-scale individual producers carry out their activities in areas of secondary vegetation. This is a challenging scenario for conservation of ecosystem services.

Tourism was recently developed in the BBR, based on an infrastructure consisting of an aquarium, campground, sculpture exhibition centres (one educational and the other for cultural promotion), two ornamental plant gardens, three diving areas, three villas or rest centres for domestic tourism, four museums, as well as seven hotels and villas for international tourism (Salmerón et al., 2018).

Research on the management of natural resources by Acosta (2010) placed emphasis on the core zone, the Gran Piedra Protected Natural Landscape (PNPGP) and the El Retiro Natural Reserve. The author observed that local actors fully use resources and services available, revealed by the diversity of species and the uses of the local biodiversity, that was the base to classify the biodiversity according to use received from rural communities at those places. This points to a close link between the worldview of these actors and the environment to meet their economic, social, domestic, aesthetic or magical-religious needs (Table 6.1).

An unequal distribution of the different types of biodiversity was noted in the three communities located in the core zone of the PNPGP. In the Gran Piedra and El Olimpo zones, all forms were reported while in El Desierto the last type was not identified. This can be attributed to the different conditions (or quality) of life given the relative geographic and social isolation due to its mountainous location with steep slopes, limited educational level (predominantly primary-secondary: 76% of the population), scarce road infrastructure and in poor condition, shortage of economic-social infrastructure (one family health clinic, an all-purpose store and a Video Hall), lack of electricity (supplied from a power plant operating only during the coffee harvest), with forestry as the main source of employment, which is characterised by low wages and poor technical training, among other causes. Unlike the other two communities which have infrastructure for tourism (motels, cabins and restaurants), which support the production of handicrafts, tourism does not play an important role in the El Desierto community.

Table 6.1 Types of agrobiodiversity and its use in the Gran Piedra Protected Natural Landscape (Acosta, 2010)

Type	Description	Uses identified
Natural	Biodiversity produced by evolution, with very low management and anthropic impact	*Anti-erosion barriers*, Domestic, Ornamental, Food, Medicinal
Exotic[a]	Non-native biodiversity, but acclimated and managed by communities and with historical and cultural values	*Ornamental*, Forestry, Agriculture, Religious, Food
Affective[a]	Biodiversity related to individual preferences or religious beliefs, such as pets, e.g., dogs, cats, fish, birds, turtles or flowering plants	*Ornamental, Pets*, Religious
Domestic[a]	Biodiversity associated with domestic use and as source of food contributing to the diet, such as species for human consumption, condiments and medicinal species	Condiments, *Utility, Food, Medicinal*
Productive	Biodiversity associated with agricultural production on farms, such as livestock or crop species	Condiments, Forestry, *Food, Medicinal*
Functional	Biodiversity used on farms by enhancing their ecological functions, such as pollinators, pest control species or mycorrhizae	Ornamental
Elaborated[a]	Biodiversity when transformed is used to obtain an ornamental or functional product with commercial value, such as handicrafts or costume jewellery	Ornamental, *Agricultural, Domestic, Religious*

[a]Type of biodiversity reported for the first time in Cuban protected areas. New reports detected during the investigation are shown in italics.

This diversity and functionality of the biodiversity coincides with what was observed in other areas, in particular in Latin America (Hernández et al., 1998, 2000; Fuentes et al., 2008; Trujillo and Correa, 2010; White-Olascoaga et al., 2013). Research conducted by Polanco et al. (2018) and Acosta et al. (2018) in 18 localities of the Reserve on the use of biodiversity by community residents in the area added knowledge to that reported by Rosete and Ricardo (2015) regarding the medicinal use of some animals. These last authors report the use of the Cuban land snail *Zachrysia guanensis* for medicinal purposes, while Acosta et al. (2018) report the use of another slug species (*Veronicella* sp.) for the specific treatment of rheumatic diseases, the common house borer (*Anobium punctatum*) to combat bronchopneumonia and for the first time, the house mouse (*Mus musculus*) to treat asthma-related conditions. These authors also observe that this agricultural ecosystem is a source of flora and fauna species for the treatment of health conditions, due to the number of species used by the local inhabitants. More than 80% of the agrobiodiversity is utilised primarily for medicinal purposes and food, as noted by Acosta et al. (2018), including sage (*Salvia officinalis*), light-blue snakeweed (*Stachytarpheta jamaicensis*), basil (*Ocimum basilicum*), key lime (*Citrus aurantifolia* and other species), black nightshade (*Solanum nigrum* Lin.), leaf-of-life (*Bryophyllum pinnatum*), rockweed (*Pilea microphylla*), guava (*Psidium guajava* L.), mango (*Mangifera indica* L.), sapote (*Manilkara zapota*), arrowroot (*Maranta arundinacea* L.), turmeric (*Curcuma* spp.) and afió (*Arracacia xanthorriza*, Bancr).

The prevalence of these uses is determined by several factors, including cultural, since this area was settled by French-Haitian in-migrants in the 17th and 18th centuries. Also, magical-religious beliefs based in Ibero-American syncretism and transculturation through different migratory processes in the area (Hammer et al., 1992). Inaccessibility plays a role, as communities are located in isolated mountainous rural areas with difficult access and poor infrastructure, including medical structures, and intermittent supply and shortage of medicines. And, finally, the Cuban government policy of promoting the development of mountain communities based on a sustainable use of their natural resources.

Not less important for the occurrence of these uses are the ecological and edaphoclimatic conditions of the area, characterised almost entirely by mesophyll evergreen forest vegetation (at altitudes between 500 and 800 m.a.s.l.) and montane rainforest (from 800 to 1,400 m.a.s.l.) (Álvarez et al., 2021).

Farmers' perception of diversity and the protected area

Management of biodiversity is closely related to farmer's perception of the surrounding natural environment, conveyed by verbalisations and subjective judgments, with a certain degree of elaboration by the residents (Núñez, 2003). Perception was measured using two dimensions (Figures 6.2 and 6.3): level of sensitivity, relating to the social actors' level of knowledge of biodiversity and awareness of the impact of biodiversity use on the natural environment, as well as

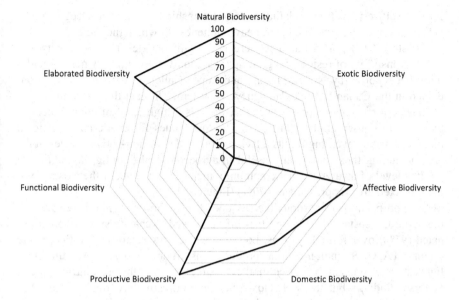

Figure 6.2 Level of perception (%) of social actors regarding biodiversity management in the PNPGP (Acosta, 2009).

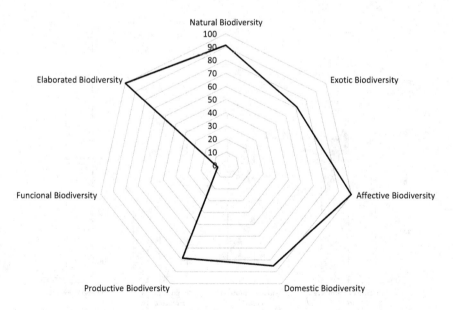

Figure 6.3 Level of self-responsibility (%) of social actors regarding biodiversity management in the PNPGP (Acosta, 2009).

level of self-responsibility relating to the relationship between knowledge and the actions taken to preserve the natural environment with which they interact.

Acosta (2010) points out that there is a significant level of sensitivity and self-responsibility of residents regarding most biodiversity types, which coincides with that reported by Núñez-Moreno and López-Calleja (2005) in research conducted in the Cuban Sabana-Camagüey ecosystem. The fact that more than 50% of these criteria are positive proves that there is a high level of commitment to biodiversity conservation in this area. The low values for exotic and functional biodiversity, however, highlights a lack of knowledge of non-native species in the area, including those potentially useful in agroecological farming, in addition to the low level of awareness that community members have about the existence of the protected area (approximately 70%) (Figure 6.4).

Seed propagation of agrobiodiversity operated by farmers is highly relevant. In this regard, Acosta (2013) found that informal seed exchange systems predominated (95%) over formal systems, both within the state sector and in the private sector (CPA, CCS, small individual producers, usufructuaries). However, this highlights shortcomings in the agricultural extension system in the area, attributable to factors including – but not limited to – relief, poor infrastructure (roads, electricity), transportation and trained personnel. There is a strong community social network, notwithstanding, that guarantees informal sector producers adequate resources for sowing and using seed supply strategies in order of importance: self-sufficiency, exchange and gifts among friends and relatives (Figure 6.5).

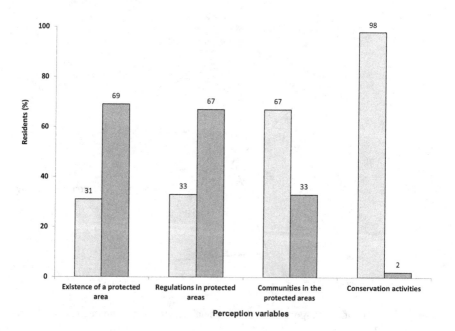

Figure 6.4 Social actors' perception of the protected area (%).

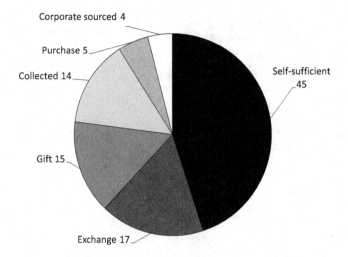

Figure 6.5 Pathways of seed exchange by farmers in the PAMR.

These results coincide with those found by the same author for farmers under different edaphoclimatic conditions (arid coastal zone) in buffer zones of the El Retiro Natural Reserve, as well as with those reported by Castiñeiras et al. (2009) in similar studies in Cuba and Mexico. This underscores the importance of community social networks as they play a fundamental role in the functioning of the informal seed system, in the resilience of the agricultural systems and in achieving local food security as well as in strengthening oral community and family traditions in agroecosystem management.

Most BBR agroecosystems, located primarily in marginal zones, mountainous and isolated, scattered areas, preserve in situ numerous cultivated species, some of which catalogued as underutilised crops and in danger of genetic erosion, e.g. the previously cited afió (Arracacia xanthorriza), sagú (Maranta arundinacea) and turmeric (Curcuma sp.), as pointed out by Rodríguez et al. (1999), and Shagarodsky and Castiñeiras (2013) in studies conducted in different areas of the country.

Acosta et al. (2012) examine the multiplicity of the uses of these species, as this region was one of the main settlement areas of French-Haitian migration in Cuba. These immigrants found favourable soil and climatic conditions for the cultivation of some species, such as the farinaceous that traditionally have been cultivated and used in various forms from one generation to another, in food preparation (fried, atol, sweet, pickled, condiment, *vianda* and thickener for stews), as medicine (poultice, diuretic and eupeptic), agronomic (living fences) and aesthetic (ornamental) (Figure 6.6).

However, these authors (Acosta et al., 2012) also point out that these species are increasingly scarce (in use and traditional cultivation) in most agroecosystems, which increases the risk of genetic erosion. Other causes contributing to this loss are the low commercial value attributed (agree 83% of those interviewed), rural migration (81%), declining knowledge of their use and cultivation, nutritional properties, as well as their ecological importance and for local food security (67%).

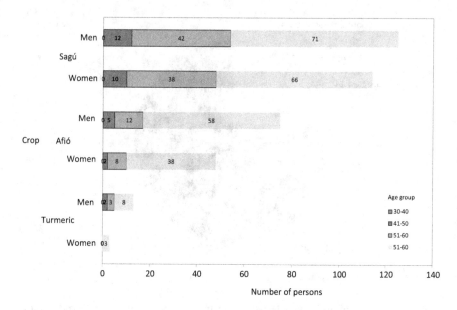

Figure 6.6 Traditional uses of *sagú*, *afió* and turmeric in PAMR areas.

These factors have led to a loss of knowledge about the three species – a legacy of the French-Haitian migration – conserved mostly by older farmers (51–60 years).

The role of women in traditional agriculture in the BBR

The gender ratio in the BBR settlements is very unbalanced, since less than 10% of women participate actively in farming. They have been sidelined from leadership positions in the cooperatives and from land ownership, as all farms belong to men or to legal entities, and women play most predominantly domestic roles. This agrees with FAO (2014) in studies on the situation of women in agriculture worldwide and with the findings of García et al. (2015), in studies on the management of traditional agroecosystems in three regions of Cuba, which corroborates the leading role of men in decision making in farming.

Acosta (2016) studied the management of rural agricultural systems and the role of women in two municipalities in the province of Santiago de Cuba (Segundo Frente and San Luis) and concludes that the status of women in the agricultural sector is determined by a patriarchal and conservative culture still prevalent in Cuban society, even more deeply-rooted in rural mountain areas. In most cases, women are in charge of domestic work and home gardens, and they are decisive in the domestic use of the agrobiodiversity, especially for in home cooking. This reveals the gap between social practice and a legal framework which guarantees equal rights and opportunities as established both by the Constitution of the Republic of Cuba and by Law 116 Labour Code (Gaceta Oficial de la República de Cuba, 2014).

Agricultural production on family and local scales

Acosta (2014) examined the destination of food production of six farmers in the community of Verraco (buffer zone of El Retiro Nature Reserve), within the framework of the international Caribbean Biological Corridor (CBC) project. This study showed that 96% of agricultural production is consumed locally, with 57% for domestic consumption, 23% for marketing (or distribution) within the community, and 16% for local markets. The remaining production, such as cherries (*Prunas* sp.) and flowers (lilies, agapanthus and others) is mainly destined for state-owned companies. Their success lies in the biological quality, organic farming, and low production costs, which allow up to one third of prices in other markets. An advantage of family farming is the closeness between production areas and local markets thus offering fresh produce and reduced costs for transportation, intermediaries and packaging. Diversification of production may lead to the commercialisation of products beyond those with food value, such as ornamental flowers (lilies, agapanthus and other).

Abdulai and Kuhlgatz (2011) observed that the local food security is based on seasonal availability to guarantee the community a secure and stable source of agricultural products and in accordance with the food culture of the area, as well as access to these products on a permanent basis in developing countries, like Cuba.

An increase in small-scale food processing has been observed, since more than 50% of the producers process their products (pickles, purees, jams, wines and preparations to alleviate certain health conditions) using low technology, for self-consumption and to a lesser extent, for sale.

The social and environmental responsibility of producers contributes to the predominance of the rural family farming production system over the cooperative system in this locality. It was observed that the production of the six farms in the study conducted by Acosta (2014) satisfies a part of the school food diet in the community (occasional donations for school snacks) and supplied – either free-of-charge or at favourable prices – agricultural products to community families in need, while also reforesting and planting fruit trees in their farm or in the community. These actions stand out above those carried out by the cooperative sector, whose production, although with a high social significance, uses channels of commercialisation that do not always benefit the local community.

References

Abdulai, A. and Kuhlgatz, C. (2011) 'Food security policy in developing countries', in J.L. Lusk, J. Roosen and J.F. Shogren (eds) *The Oxford handbook of the economics of food consumption and policy*, Oxford University Press, pp344–369

Acosta, G. (2010) 'Percepción y manejo de la biodiversidad por los actores sociales del paisaje natural protegido Gran Piedra en Santiago de Cuba', Master thesis on Agrarian Extension, Universidad Agraria de La Habana, La Habana, 52pp

Acosta, G. (2013) 'Evaluación del sistema de abastecimiento de semillas en agroecosistemas tradicionales del Paisaje Natural Protegido Gran Piedra', *Agricultura Orgánica*, vol 1, pp7–11

Acosta, G. (2014) 'Agricultura familiar campesina y satisfacción nutricional en la comunidad de Verraco', *Leisa Revista de Agroecología*, vol 30, no 4, pp22–24

Acosta, G. (2016) 'Resultados del análisis de sistemas agrícolas en comunidades de Segundo Frente y San Luis (La Caoba), Provincia Santiago de Cuba, Informe final de resultados', Centro Oriental de Ecosistemas y Biodiversidad (Bioeco), Santiago de Cuba, 14pp

Acosta, G., Abad, M.A., Ramos, L., Izquierdo, J.E. and La Llave, S. (2012) 'Manejo agrícola participativo de especies en peligro de erosión genética, en el Paisaje Natural Protegido Gran Piedra', Centro Oriental de Ecosistemas y Biodiversidad (Bioeco), Santiago de Cuba, 15pp

Acosta, G., Polanco, G., Figueredo, L.M., Garay, L. and Pérez, N. (2018) 'Usos medicinales de los recursos de la biodiversidad en comunidades de la reserva de la biósfera de Baconao', *Agrotecnia de Cuba*, vol 42, no 1, pp35–48

Álvarez, L.O. (2022) 'Localización de la Reserva de la Biosfera Baconao', personal communication

Álvarez, L.O., Salmerón, A., Acosta, G., Fagilde, M.C., Álvarez, F., Silot, M., Abad, M.A., Fong, A., Costa, J. and Sanfiel, M. (2021) 'Plan de Manejo para el paisaje natural protegido Gran Piedra. Periodo de manejo 2022-2026', Centro Oriental de Ecosistemas y Biodiversidad (Bioeco)/Empresa Agroforestal Gran Piedra-Baconao, Santiago de Cuba, 366pp

Castiñeiras, L., Cristóbal, R., Pinedo, R., Collado, L. and Arias, L. (2009) 'Redes de abastecimiento de semillas y limitaciones que enfrenta el sistema informal', in M. Herman, K. Amaya, L. Latournerie and L. Castiñeiras (eds) *¿Cómo conservan los agricultores sus semillas en el trópico húmedo de Cuba, México y Perú? Experiencias de un Proyecto de Investigación en Sistemas Informales de Semillas de Chile, Frijoles y Maíz*, Bioversity International, Rome, pp73–84

CNAP (2013) 'Plan del Sistema Nacional de Áreas Protegidas 2014-2020', Centro Nacional de Áreas Protegidas e Instituto de Investigaciones Fundamentales en Agricultura Tropical Alejandro de Humboldt, Ministerio de Ciencia, Tecnología y Medio Ambiente, La Habana, 365pp

FAO (2014) 'Política de igualdad de género de la FAO', Food and Agricultural Organization, http://www.fao.org/docrep/019/i3578s/i3578s.pdf, accessed June 6 2022

Fuentes, V., Cristóbal, R., Shagarodsky, T., Castiñeiras, L., Fundora, M., Barrios, O., Moreno, V., Fernández, L., Orellana, O., Alonso, L., González, V., García, M., Giraudy, C., Valiente, A. and Hernández, F. (2008) 'Plantas ornamentales en conucos de tres regiones de Cuba', *Plant Genetic Resources Newsletter*, no 140, pp51–56

Gaceta Oficial de la República de Cuba (2014) *Gaceta Oficial No. 29 Extraordinaria de 17 de junio de 2014*, https://www.gacetaoficial.gob.cu/sites/default/files/go_x_029_2014.pdf, accessed December 15, 2021

García, M., Castiñeiras, L. and Bonet, A. (2015) 'Biodiversidad en huertos caseros y fincas de Cuba', in S. Rosete Blandariz and N. Ricardo Nápoles (eds) *Biodiversidad, usos tradicionales y conservación de los productos forestales no maderables en Cuba*, Universitat D'Alicant, Alicante, pp235–266

González, H., Rodríguez, L., Rodríguez, A., Mancina, C.A. and Ramos, I. (2012) *Libro rojo de los vertebrados de Cuba*, Instituto de Ecología y Sistemática, ARG Impresores, Madrid, 304pp

Hammer, K., Esquivel, M. and Knüpffer, H. (1992) *Origin, evolution and diversity of Cuban plant genetic resources*, Institut für Pflanzengenetik und Kulturpflanzenforschung, vol 1, Gatersleben

Hernández, J., Ramos, P.E. and Acosta, F. (1998) 'Usos de las plantas', in N. Viña (ed) *Informe final del proyecto Diversidad Biológica del Macizo Montañoso Nipe-Sagua-Baracoa*,

Tomo III, Centro Oriental de Ecosistemas y Biodiversidad (Bioeco), Santiago de Cuba, pp678–749

Hernández, J., Temó, D. and Acosta, F. (2000) 'Usos de las plantas', in N. Viña *Informe final del proyecto Diversidad Biológica del Macizo Montañoso Sierra Maestra*, Tomo I, Centro Oriental de Ecosistemas y Biodiversidad (Bioeco), Santiago de Cuba, pp459–503

IUCN (2019) *The IUCN Red List of Threatened Species*, Version 2019-2, https://www. iucnredlist.org, accessed September 11, 2021

Núñez, M.L. (2003) 'Sostenibilidad y actores sociales en la protección del medio ambiente en Cuba', V Encuentro de Política Social y Trabajo Social, San José, http://biblioteca.clacso. edu.ar/ar/libros/cuba/cips/caudales05/Caudales/ARTICULOS/ArticulosPDF/0416N076. pdf, accessed March 7, 2022

Núñez-Moreno, L. and López-Calleja, C. (2005) 'El medio ambiente y la biodiversidad en las percepciones de comunidades costeras cubanas. Informe del Centro de Investigaciones Psicológicas y Sociológicas', in P.M. Alcolado, E.E. García and M. Arellano (eds) *Impactos, lecciones aprendidas y desafíos del Proyecto Ecosistema Sabana-Camagüey. Estado actual, avances y desafíos en la protección y uso sostenible de la biodiversidad*, Editorial Academia, La Habana, pp170–175

Polanco, G., Figueredo, L.M., Almarales, A., Revilla, Y., Castell, M.Á. and Baró, Y. (2018) 'Componentes de la diversidad biológica empleados para las familias cubanas en la medicina natural y tradicional', Proyecto Nacional del Programa Medicina Natural y Tradicional, Centro Oriental de Ecosistemas y Biodiversidad (Bioeco), Santiago de Cuba, 57pp

Rodríguez, A., Rodríguez, N.A. and Quintero, F.S. (1999) 'Caracterización de Germoplasma y Mejoramiento Participativo en Especies de Raíces y Tubérculos Alimenticios Tropicales y en Musáceas', *Resúmenes Simposio Internacional y Talleres sobre Fitomejoramiento Participativo (FMP) en América Latina y el Caribe: Un Intercambio de Experiencias*, Ecuador, pp1–13, http://ciat-library.ciat.cgiar.org/FitomejoramientoParticipativo/ html/inicio.htm, accessed November 14 2022

Rosete, S. and Ricardo, N.E. (2015) *Biodiversidad, usos tradicionales y conservación de los productos forestales no maderables en Cuba*, Universitat D´Alicant, Alicante, 280pp

Salmerón, A. and Álvarez, L.O. (2014) 'Plan de manejo Área Protegida de Recursos Manejados Reserva de Biosfera Baconao', Centro Oriental de Ecosistemas y Biodiversidad (Bioeco), Santiago de Cuba, 114 pp

Salmerón, A., Acosta, G., Silot, M., Fagilde, M.C., Álvarez, L. O. and Abad, M.A. (2018) 'Plan de Manejo para el Área Protegida de Recursos Manejados Reserva de Biosfera Baconao - Periodo 2019 a 2023', Centro Oriental de Ecosistemas y Biodiversidad (Bioeco), Santiago de Cuba, 25pp

Shagarodsky, T. and Castiñeiras, L. (2013) 'Especies de plantas subutilizadas en Cuba', *Agrotecnia de Cuba*, vol 37, no 1, pp18–25

Trujillo, W. and Correa, M. (2010) 'Plantas usadas por una comunidad indígena Coreguaje en la Amazonía colombiana', *Caldasia*, vol 32, no 1, pp1–20

White-Olascoaga, L., Juan-Pérez, J.I., Chávez-Mejía, C. and Gutiérrez-Cedillo, J.G. (2013) 'Flora medicinal en San Nicolás, municipio de Malinalco, Estado de México', *Polibotánica*, no 35, pp173–206

7 Nature's Matrix

The quality of managed and natural landscapes

Ivette Perfecto and John Vandermeer

Introduction

Few would deny that Capitalism has had negative effects on the majority of the world's human population, even as it has created wealth for the few. Even fewer would deny that Capitalism has had negative effects on the natural world. From the unbridled and unsustainable extraction of natural resources to the pollution of the very nature that is needed to sustain life, the natural world is nothing but an opportunity for extracting yet more surplus value to sustain the capitalist venture (O'Connor, 1994; Harvey, 1996). The human suffering is immense, yet Capitalism's defenders point to the obvious benefits already reaped, suggesting that the oppression and misery of many is somehow thus justified. Searching for a similar argument with regard to the natural world is largely a fool's errand. There seems to be no silver lining. Such is especially evident in the current international concern about the ongoing loss of species and ecosystems. The international community of conservationists continually cry the alarm, noting that we currently face what some call the sixth mass extinction (Kolbert, 2014). It is clear, argue these conservationists, that we need to organize ourselves in such a way that this ongoing disaster comes to an end.

Schemes to adjust human societies in a way that would curtail this ongoing crisis are abundant in the theoretical ramblings of the international conservation community. Exemplifying both the extreme nature these ramblings can sometimes take as well as their disconnect with the scientific consensuses of the day, the late E. O. Wilson (2016) in his book entitled *Half Earth: The Planet's Only Chance*, argued that as much of one half of the planet's terrestrial surface needs to be declared "preserved" (by which he means, no human activity other than scientific study), in order to avoid the catastrophe of species extinctions. Then he adds that in the other half we should use the latest technologies to produce as much as possible, in what appears to be a strong support for current industrial agriculture. As unrealistic as his vision may be politically, what is more important is that it is much at odds with what we understand scientifically about biodiversity. For example, a recent study in Germany noted that within nature protection areas there has been a 76% decline in flying insect biomass in 27 years (Hallmann et al., 2017). Another example is the dramatic extinction rates observable in North American national

DOI: 10.4324/9781315183886-8

parks – even the largest ones (Newmark, 1995). Such observations concord well with the understanding that modern ecology brings to bare on the issue. Reserve areas occur within a matrix of managed areas and if the "quality" of that managed area is low, the effect on the reserves themselves is profoundly negative. The "half earth" philosophy mistakenly thinks that partitioning the world into biodiversity-friendly places and biodiversity-unfriendly places is possible. As we have argued in our book *Nature's Matrix* (Perfecto et al., 2009), and repeat in this chapter, biodiversity conservation is inevitably a landscape problem. One cannot allow industrial agricultural activity to go unchecked in one area and expect it to have no effects on other areas. Indeed, if we were to write a book about biodiversity conservation the title might be "Whole Earth: the planet's only chance."

Much of the conservation literature maintains the religious-like assumption that Capitalism must remain a dominant form of political organization world-wide. From an environmental perspective, this inevitably involves a contradiction between supplying human needs through the vehicle of surplus value extraction and the anti-capitalist vision of environmental protection in general, part of which is the conservation of biodiversity. In the rural sector this contradiction frequently takes the form of, on the one hand, the need to conserve natural landscapes to avoid extinctions of species and ecosystems, and, on the other hand, the need to modify those landscapes to produce surplus value and, indirectly, food and other products to meet the needs of people. In a socialist country, at least in principle, this basic contradiction is not an inevitability. Freed from the need to extract surplus value from nature and labor, Cuba is in a position to reimagine the conservation versus use value contradiction that emerges elsewhere, but in some ways continues to threaten Cuba, despite its rejection of the Capitalist system. In this chapter, we examine this dilemma in the special context of Cuban agriculture and biodiversity conservation and the COBARB (Conservation of Agricultural Biodiversity in the Biosphere Reserves of Cuba) Project. We propose a matrix approach to reconcile both of the urgent needs of humanity, biodiversity conservation, and agricultural production.

Nature's Matrix: a paradigm from ecological theory

Very few places on Earth have been completely untouched by *Homo sapiens* (Ellis et al., 2021), and there is no reason to believe that its current state is somehow eternal, as we know from many fossil sequences. The so-called pristine rain forests of the Amazon basin, for example, are riddled with current and anthropological evidence of massive human modification of the landscapes in the recent past, apparently fostering intensive soil management for intensive traditional indigenous agriculture (Denevan, 1992; Iriarte, 2007; Clement et al., 2015). Palynological studies of the eastern forest biome of North America show dramatic shifts in species composition, strongly suggesting that the famous "climax forests" described by early community ecologists is likely a temporary state after all, calling into question the entire idea of pristineness. Early ecologists used to talk about the "Oak/Hickory climax" as one of the major pristine forest types of eastern North

America, yet we now know it is a temporary vegetation formation resulting from Native American hunting and agricultural management (Abrams, 1998), gaining pristine status through the need for White colonists to claim a "wilderness" devoid of real people (Cronon, 1996). Indeed, the most recent evidence, compiled with up to date archeological and paleoecological methods for reconstructing land use reveals that 12,000 years ago nearly three quarters of the Earth's land was 0ccupied and managed by human societies, including more than 95% of temperate and 90% of tropical woodlands (Ellis et al., 2021).

The last 50 years of ecological science have produced a certain degree of agreement among ecological scientists. Various theories are relevant to the arguments covered in this chapter, most notably the theory of island biogeography (MacArthur and Wilson, 1967) and the theory of metapopulations (Levins, 1969). The later has been quite useful in articulating the problems with managing natural landscapes, particularly from the point of view of avoiding extinction. Metapopulation theory, pioneered by the late Richard Levins (1969), notes that most of nature occurs as a patchwork of habitats, some of which may be occupied by a particular species, others of which may be absent that species. This position is contrary to the previously held position that, at least as a first approximation, all populations could be viewed as uniformly spread out over the landscape and vital rates (reproductive rates, death rates, growth rates, etc.) could be taken as average over the whole population. Metapopulation theory, in contrast, recognized that most populations occur in a mixture of discrete habitats, some of which are conducive to a species' survival, others of which may be inimical thereof. Taking the simplest case, Levins postulated two patch types, habitable and uninhabitable. Then, rather than being concerned with classical vital rates, he postulated that the movement of species among patches would eventually lead to all occupiable patches being occupied. But the countervailing dynamic is that local extinctions are natural and inevitable and would ultimately counter the tendency for a species to occupy all available habitats. Balancing movement, in some cases called migration, from patch to patch, with extinction from those patches, one could easily see the way a balance was forged.

Nature's Matrix in the context of the land sparing/land sharing debate

Many recent analysts argue that we must categorize the land into preserved versus non-preserved (the latter of which consists mainly of agriculture, pastures, and plantation forestry). The non-preserved area presumably functions to provide *Homo sapiens* with what it needs, so if we make it as productive as possible, we need less of it and thus can devote more to preserved area. That is, we can spare more land for nature conservation if we intensify production on the non-spared land. This is the position argued by E. O. Wilson (2016) in his book *Half Earth*. Other analysts take a position closer to the other extreme and view the non-preserved habitat as an additional important repository of biodiversity to be shared with productive activities. This debate between "land sparing" and "land sharing"

(Perfecto and Vandermeer, 2010; Phalan et al., 2011) is similar to previous attempts at simplifying the problem of conservation [e.g., the Single Large or Several Small (SLOSS) debate (Lindenmayer et al., 2015); or the Forest Transition Model (FTM) (Mather, 1992); or Integrated Conservation Units (ICUs) (Berkes, 2004)], and similarly is, in our view, a poor formulation of the problem in the first place (Perfecto and Vandermeer, 2010).

Our alternative, popularly known as the "quality of the matrix" (Perfecto et al., 2009), might be termed "Whole earth: our planet's only chance," as we noted above. It acknowledges that all ecosystems are effectively open systems, with species coming and going, surviving for a while in some local places, but eventually going extinct, only to be recolonized from other areas in the future. This dynamic occurs whether we are talking about fragments of "natural habitat" or extensive large areas of "natural habitat." The problem is not what is in those fragments or large areas, but what is in between. That is, biodiversity dynamics operates at a large landscape level with some patches of the landscape amenable to some species but not others, other patches useful not as perennial repositories of particular populations but as temporary waystations (sometimes referred to as propagating sinks; Vandermeer et al., 2010) for species to migrate from one patch to another. The point is that the focus of conservation needs to include the entire landscape. Allowing for a few refuges in the middle of a biological desert (which is what most of industrial agriculture has become), may make romantic conservationists feel as though some conservation is happening, but such a position, while politically convenient, is a biological disaster. In Germany, the loss of 76% of insect biomass in 27 years occurred inside reserves scattered within landscapes of intensive agriculture (Hallmann et al., 2017). Extinctions will build up from local to regional, and without the continual local replenishment from migration, will turn into global extinctions. It is precisely the reason we vaccinate people for diseases – we reduce the migration (transmission) potential of the disease organism from fragment to fragment (person to person) and, by reducing the migratory (transmission) potential of the disease organism, we make it go locally extinct (stop the epidemic). We presume it is obvious that this epidemiological strategy is not something we should be trying to do in the case of conservation. Furthermore, the strategy of separating humans from nature to protect the later, has not worked as intended. A recent study that examined more than 12,000 protected areas across 152 countries concluded that protected areas failed to prevent human pressure (Geldmann et al., 2019). Indeed, in tropical regions conversion of forest to agriculture was higher in protected areas than in matched unprotected areas.

Natures Matrix in the context of the Cuban reality and the Biosphere Reserves (BR)

In the particular context of Cuba, six large Biosphere Reserves (BR) have been established with relatively standard conservation goals in mind. But basic metapopulation arguments apply here as well. BR are effectively natural areas embedded within a matrix of managed and human altered systems (agriculture, pastures, plantation forests, urban and peri-urban areas, etc.). While the BR themselves have been set up to minimize extinction rates, they too must be thought of in terms of the matrix surrounding them. They are, in the end, fragments of natural vegetation in a sea of agriculture and other human activities.

The transformation of some of Cuba's agriculture from large scale intensive state farms to smaller scale diverse agroecological farms (Funes et al., 2002; Funes-Monzote et al., 2009; Rosset et al., 2011; Altieri and Funes-Monzote, 2012; Altieri et al., 2012) is likely to contribute to increasing the quality of the agricultural matrix in Cuba as a whole. In particular, the diversification of the farms and the reduction or elimination of pesticides are likely making the agricultural matrix more permeable and allowing the movement of organisms between the large areas of the BR and all the other conservation areas in the country. Agroforestry systems, in particular, contribute to increasing the quality of the matrix in areas where the fragments of natural areas are forests. Likewise, the expansion of urban and peri-urban agroecological farms could contribute to increasing the quality of the matrix in highly urbanized areas.

Within the BR themselves, the matrix quality argument still applies but in areas where the dominant landscape is composed of forest with small patches of agriculture, the argument is effectively inverted. Those populations that carry out their vital biological functions on the isolated farms do interact with the matrix, which in this case is the "natural" vegetation that surrounds them. Thus, for example, parasitic wasps that may act as biological control agents on the farm, may very well seek their floral nourishment from the surrounding vegetation, where flowers are abundant.

The COBARB project has made significant progress assessing the existent agro-biodiversity within the Sierra del Rosario Biological Reserve in Artemisa and the Cuchillas del Toa Biological Reserve in Guantánamo. The project has also promoted the diversification of the agricultural systems within the BR by supporting seed banks and seed exchanges (Figure 7.1). Most of the farms within the reserves are too small to represent a barrier to the movement of most wildlife within the BR. However, if these areas are intensified, converted to monocultures, and expanded in area, they could have a detrimental effect on wildlife through spill-over effects (Tscharntke et al., 2005), as has happened across the tropical capitalist world.

The COBARB project's focus on these two BR presents us with a unique take on the general subject of the integration of biodiversity and agriculture. The two BR were formally established with the goal of biodiversity conservation coupled with the agricultural production and other human activities already extant in the reserves. That is, there is no formal recognition of agricultural areas and "natural" areas, but rather the system is considered as a complicated interacting system where small-scale farmers, practicing mainly traditional methods (which generally could be classified as agroecological) are interspersed within the "natural" areas. The goal of the project is to elaborate how agricultural practices of the farming families living within the reserves impact biodiversity, and how the wildlife within the reserve impacts the agricultural systems. In short, from the farmers' point of view, biodiversity in general, both on the farm and in the surrounding environment

Figure 7.1 A community seed bank in the Sierra del Rosario Biosphere Reserve. Insert: Bottles with a variety of seeds of local varieties of crops.

(agrobiodiversity and wild biodiversity or planned and associated biodiversity), could provide ecosystem services (e.g., wild bees pollinate many crops), but could also generate problems (herbivores from the forest become pests on the farm). Likewise, the agricultural practices of the farmers could provide conservation benefits by providing habitat for forest species or providing a high-quality agricultural matrix through which wildlife can migrate, or, on the other hand, could affect biodiversity negatively, like pesticide applications or intensive tillage and fertilizer application. In other words, rather than engage in what has become a rather sterile debate about the tradeoffs of agriculture and conservation, the project seeks to understand what are the true dynamics of a practicing system in which agriculture and the "natural" world are already dynamically integrated. This approach is in stark contrast with the now obsolete approach of separating natural areas from agricultural areas by establishing reserves and excluding people from their traditional/ancestral lands (Siurua, 2006).

The central part of both BR is rugged and penetrated only with difficulty. The vegetation is largely forest that has seen little intervention, at least since the time of the European conquest. Dotted within this central area are small farms of various sizes, perhaps 50–100 in number. These farms clearly have the most direct physical relationship with the surrounding forest (indeed, one farmer, when asked "where does the forest begin?", told us "the farm is part of the forest"; Figure 7.2a). Management of farms with this intimate contact with the forest requires acknowledgment that the farm and forest are not all that different (something the farmers themselves seem to fully appreciate).

At the other extreme, farms and/or urban settlements seem to dominate the parts of the landscape that are not within those rugged central parts. Patches of forest are

Figure 7.2 Farms within the Sierra del Rosario Biosphere Reserve in Artemisa Province. (a) An agroforestry farm inserted in a mostly forested landscape and (b) a multi-functional farm inserted in a mostly agricultural landscape.

inevitably present, but the generalized matrix is one of farms, and the "fragments" in that matrix are small patches of forest, usually secondary with evident human use (Figure 7.2b). Here the farmers are less concerned with the native biodiversity but must be concerned with the on-farm biodiversity, especially with regard to the provisioning of ecosystem services. For example, if the biodiversity of bees that maintain adequate pollination services becomes reduced by the reduction of floral biodiversity in forest patches, that is of considerable concern. At this extreme, where agriculture is common in the landscape, the effect of the farm on the biodiversity within the reserve becomes a significant issue. If the migratory pathways in a high-quality matrix become disrupted by intensive monocultural farm patches in that matrix, the ultimate effect on natural biodiversity could be negative.

In sum, the situation can be viewed as a continuum in which small farms are isolated one from another and located in a matrix of natural forest at one extreme, to the situation in which small patches of natural forest are isolated from one another and located in a matrix of agriculture (Figure 7.2). The sorts of scientific questions about biodiversity are bound to be distinct depending on where a farm or forest is located along that continuum.

It is also important to recognize that the goals of BR are not only to conserve biodiversity but also to contribute to the livelihoods of the families living within them and in the surrounding areas (Batisse, 1986). To reconcile these two objectives the COBARB project has assessed environmental impacts of some agricultural practices within the Reserves and promoted alternatives to mitigate such impacts. Agrobiodiversity can play a major role in wildlife conservation and supporting farmer's livelihoods. Not only do diverse farming systems promote functional diversity, such as pollinators and natural enemies, thus contributing to the productivity and sustainability of the farms, they also represent sources of income for the farmers living within the reserve. The COBARB project has explored potential markets for a variety of non-traditional crops, to support the maintenance of these diverse farming systems within the reserves. Another potential to be explored within the biosphere reserve is the connection between farming and tourism, such

Figure 7.3 A farmer in the Sierra del Rosario Biosphere Reserve showing the trap that he uses for the control of the coffee berry borer. In the background is visible the coffee agroforestry system that he manages with diverse nitrogen fixing trees. These are practices that he described as agroecological.

as the cultivation of crops that could be used for crafts or to supply produce and products to hotels and restaurants within the reserve.

The support of agroecological practices within the reserves is also essential to maintain the integrity of these conservation areas (Figure 7.3). Pesticide applications have been shown to be highly detrimental to biodiversity and to have spillover and indirect effects on wildlife. As mentioned earlier, the study in Germany found a 76% decline in insect abundance in protected areas within an agricultural landscape (Hallmann et al., 2017), even though those fragments had effective and complete protected status. Furthermore, declines in bird populations have been attributed to the decline in insects and pesticide applications, at least in Europe (Hallmann et al., 2014; Silva et al., 2018). All this evidence suggests that intensive/industrial agriculture should not be allowed within the reserve areas and should be avoided even outside the BR.

The political argument

Food sovereignty and the role of small-scale family farmers

Ten years after the fall of the Berlin wall, Levins (1998: 63-64) in an article about Cuba during the Special Period entitled *Rearming the Revolution: the tasks of theory for hard times* said:

In agriculture we criticize the view that sees progress as occurring along a single pathway from small scale petty production to large scale agribusiness, from subordination to nature to control over nature and from traditional superstition to modern science.

We recognize this development pathway as driven by class struggle and the conversion of knowledge unevenly into commodities. We counterpoise an ecologically and socially rational pathway that goes beyond brute force capital intensity to knowledge-intensive low input practices, beyond the apparent efficiencies of monoculture to planned diversity in which the agricultural enterprise is a planned mosaic of fields in which each has its own product but also contributes to the productivity of the other fields. We invent ways to reduce rather than increase input, with the methods of bio-fertilizers, nitrogen-fixing microorganisms, mineral-mobilizing fungi, recycling of litter with the help of invertebrates, and natural pest management aimed at systems that are as self-operating as possible. We see both peasant and scientific knowledge as contributing to the design of these systems. Cuba has adopted this pathway of development partly out of conviction and partly from necessity in the special period, so that the commitment is very real and powerful but not yet consolidated. Finally, we note that such a program is contrary to the capitalist need for technologies that require more, not less, investment per hectare.

Today, although the commitment to agroecology is still not consolidated in Cuba, agroecology has become a movement that is strongly supported by, among others, the ANAP (National Association of Small Scale Farmers), a member of the international peasant farmer organization, La Via Campesina (Machín Sosa et al., 2010; Rosset et al., 2011). At approximately the same time Levins was writing the previous words, La Via Campesina coined the term "Food Sovereignty" after lengthy debate on how to incorporate the various ideas and agendas of their extremely broad-based membership into a single identifier (Martínez-Torres and Rosset, 2014), effectively accepting Levins' call to "weave together" distinct progressive takes on the food and agriculture system. The concept of food sovereignty goes beyond "food security," acknowledging the fundamental right of local communities to decide what should be produced, how, for whom, and by whom. The antiquated notion that simple equilibrium capitalist rules should be a foundation for food production is explicitly rejected, and the needs of "the people" should be that foundation. The idea incorporates the rights of small-scale farmers to produce what they want and how they want in the context of the needs and desires of local communities, and effectively strengthens local food systems. Intrinsic to the concept of food sovereignty is the autonomy of farmers, whether from large corporations or the state, and this is where agroecology plays an important role. Farming agroecologically offers the opportunity for higher levels of autonomy since agroecological systems rely on biodiversity for providing ecosystem services such as pollination, pest control, nutrient cycling, etc. Also, one of the principles of food sovereignty includes de conservation of nature, which implies conservation of biodiversity. Here we see the intricate relationship between food sovereignty, diverse agroecological systems and the conservation of biodiversity (Chappell et al., 2013). It is this intersection that the COBARB project emphasized within the BR.

Chayanovian balances in the context of food sovereignty, agroecology, and biodiversity

At the center of the intersection between food sovereignty, agroecology, and bio-diversity is the farmer (Perfecto et al., 2009). Farmers are the ones making deci-sions that change the landscape, in good and bad ways. To understand and design agricultural systems that can produce food and conserve biodiversity it is impera-tive that we understand not only the ecological aspects of the relationship between food production and biodiversity conservation, but also the social, political and economic aspects. A socio-ecological approach allows for a better understanding of how the physical and social environment influence farmers to do what they do, individually or collectively. The agriculture found within the BR in Cuba is domi-nated by small-scale farmers and therefore, it is important to understand how these farmers make decisions.

Rural sociologists and agricultural economists have been debating the peasant condition for over 100 years, with both Marxists and capitalist economists pre-dicting the demise of the peasantry (van der Ploeg, 2013). But, rather than dis-appearing, today there are notable examples of repeasantization throughout the world (van der Ploeg, 2013). In Cuba, the small farmer sector increased with the dissolution of some of the state farms during the Special Period (Machín Sosa et al., 2010). Part of the growth of the peasantry globally has to do with population growth, but the persistence of the peasantry, in spite of the harsh economic condi-tions they are forced to endure, is testimony to their resistance and their constant search for autonomy (van der Ploeg, 2018). In order to understand how today's peasant and small-scale farmers engage in this resistance, rural sociologist Jan Douwe van der Ploeg expanded on the extensive field work of Russian agrarian economist Alexander V. Chayanov (1925). Chayanov engaged in detailed study of the Russian peasantry of the first quarter if the nineteenth century. Contrary to both the super atomized theories of the capitalist economists or the class-based theories of Marx, Engels, and Lenin, Chayanov found that it made little sense to think of the Russian peasant as an atomized individual seeking to maximize utility. Indeed, the very notion of encapsulating all human activity into the idea of utility did not seem nearly as useful as it may have been when applied to the factory worker. Rather, according to Chayanov, the small-scale farmer is constantly adjudicating a series of "balances." In planning for grain production, for example, the farmer seeks to balance the need of the family for food with the need for generating money to purchase other goods; seeks to balance the utility of a new technology with the drudgery of applying it; seeks to balance the responsibility of social contract (e.g., taxation) with the responsibility of maintaining familial social security; and many other balances. This is to say that the farmer is constantly and simultane-ously thinking of all of these balances as she plans the productive activities of the farm. van der Ploeg (2018) actualized Chayanov's analysis by adding further evident balances that are more relevant to contemporary peasants. Encapsulating this complex of balances into a single idea of "utility" is certainly possible, but not all that useful. It is, however, possible to compare the fundamental capitalist mode, in which maximization of utility seeks to encapsulate the bulk of human behavior (at least to a first approximation) to the Chayanovian mode in which balances are continuously adjudicated with shifting attention, perhaps even from moment to moment. At one end of the spectrum is the farmer who specializes on a specific commodity that enters a well-oiled market system, what van der Ploeg (2018) refers to as the "entrepreneurial farmer[1]." At the other end of the spectrum is the farmer with minimal penetration into capitalist markets, juggling a plethora of balances that determine what is to be done on the farm, and therefore maintain a diverse multifunctional system to satisfy a variety of needs. A given farm, whether in nineteenth century Russia or 21st-century Cuba, can be quantitatively thought of as on a point along a continuum that ranges from 100% Chayanovian to 100% capitalist. Of course, in the context of the Cuban economy, no farmer will fall in the "100% capitalist" category. Nevertheless, a specialized state farm that focuses on one or a few crops, is fully inserted in the market, depends on external inputs to maintain production, and strives for financial efficiency, could be considered as parallel to the "entrepreneurial farmer," and be usefully placed at the "100% capitalist" extreme of the continuum.

In a visit that we made to the Sierra del Rosario Biosphere Reserve, we met several peasant families, each of whom produced at least 20 different crops and some small animals. Some of them had very diverse agroforestry systems. For these farmers, these multifunctional systems represented a way to survive in a harsh ecological and economic environment. The diversity of crops resulted in a diversification of their food but also a diversification of sources of income through local markets (Figure 7.4). Based on how inserted they are into external markets and based on their management decisions of financial efficiencies, these families could be placed on the left side of the Chayanovian-Capitalist gradient (Vandermeer and Perfecto, 2012; Valencia Mestre et al., 2018). At the other extreme, we could place, for example, a state conventional farm that produces oranges for the export market. We can imagine such farms producing oranges in monocultures, with agrochemical inputs, a hired labor force, and with a production that goes entirely to the market. Through this type of analysis we can begin to see the interpenetration of environmental, socio-economic, and political forces that move farmers to do what they do and how that relates to biodiversity conservation. Although such socio-political issues require further thought and study, especially in the context of a socialist economy, it is possible to see the connection between a peasant family striving for autonomy and the decision to produce agroecologically.

It is important to emphasize that the connection between food sovereignty, agroecology, and biodiversity is not something that emerged from the centers of academia. Rather, the analysis first emerged from the peasant movement itself, where traditional and local knowledge intersects with the natural world (Toledo, 2005; Martínez-Torres and Rosset, 2014). However, both, academic agroecologists and farmers practicing agroecology agree that the fundamental rules of natural systems, which is to say ecology, should be used as guidelines for planning and management.

Figure 7.4 A farmer in the Sierra del Rosario Biosphere Reserve showing the multipurpose salve that she developed with more than 30 plant ingredients, all of which she grows on her farm. Insert: bottled salve with a list of the ingredients.

Conclusion

We conclude by citing Richard Levins again:

> … some [people] … analyze the production system as an ecosystem, … others look at the structure of injustice … and still others … examine the demography of rural [and urban] areas, these are not different opinions about the same thing, but [rather] different agendas. They are all clearly legitimate for their own purposes [yet] … incomplete … [with respect to] understanding the whole. The next step then is to weave them together, to show that they are partial, relatively true perspectives on a greater whole, and to examine how class structure or land tenure affect the technology of production which in turn alters the ecology, etc. In the traditional use of the story about blind men and an elephant, they each describe the elephant from their partial perspectives. The story is used to conclude that each one has his own reality. But another interpretation is possible: there really are elephants; and the men should talk to each other. If the story were retold as four blind women, they would put together, by communicating, a more comprehensive view of elephants!

The COBARB project in Cuba is attempting to put together a more comprehensive view of agrobiodiversity within the BR in Cuba. The objective of conserving, managing, and enhancing the use and sharing of agrobiodiversity (crop varieties, wild relatives, useful plants, etc.) to improve the livelihoods of the farmers that nurture such agrobiodiversity within the MaB (Man and Biosphere) reserves can be accomplished only with this comprehensive approach. By rejecting the old notion of the separation of nature and people and insisting in an integrative approach, the COBARB project is in the vanguard of conservation and socially just development. Conservationists and agricultural experts from the rest of the world could learn from this example.

Note

1 van der Ploeg (2018) identifies one extreme as the capitalists/corporate farms which are part of Empire, as defined by Hardt and Negri (2000). However, in the context of the Cuban's socialist economy, this sector does not exist. The other type of farmers identified by van der Ploeg is the "entrepreneurial farmer" who is fully inserted in the market, specializes on one or very few crops, depend on external inputs to maintain production and has profit maximization as its main objective. In the context of Cuba's economy, a large state farm may resemble more these "farmers" with the exception of the profit maximization objective.

References

Abrams, M.D. (1998) 'The red maple paradox', *BioScience*, vol 48, no 5, pp355–364
Altieri, M.A. and Funes-Monzote, F.R. (2012) 'The paradox of Cuban agriculture', *Monthly Review*, vol 63, no 8, pp23–33
Altieri, M.A., Funes-Monzote, F.R. and Petersen, P. (2012) 'Agroecologically efficient agricultural systems for smallholder farmers: contributions to food sovereignty', *Agronomy for Sustainable Development*, vol 32, no 1, pp1–13
Batisse, M. (1986) 'Developing and focusing the biosphere reserve concept', *Perspectives in Resource Management in Developing Countries*, no 1, pp160–177
Berkes, F. (2004) 'Rethinking community-based conservation', *Conservation Biology*, vol 18, no 3, pp621–630
Chappell, M.J., Wittman, H., Bacon, C.M., Ferguson, B.G., Barrios, L.G., Barrios, R.G., Jaffee, D., Lima, J., Méndez, V.E., Morales, H., Soto-Pinto, L., Vandermeer, J. and Perfecto, I. (2013) 'Food sovereignty: an alternative paradigm for poverty reduction and biodiversity conservation in Latin America', *F1000Res*, vol 2, 235. https://pubmed.ncbi.nlm.nih.gov/24555109/, accessed November 6 2022
Chayanov, A.V. (1925) *The theory of peasant economy*. Edited by D. Throner, B. Kerblay and R. Smith; with a foreword by T. Shanin (1986), University of Wisconsin Press, Madison
Clement, C.R., Denevan, W.M., Heckenberger, M.J., Junqueira, A.B., Neves, E.G., Teixeira, W.G. and Woods, W.I. (2015) 'The domestication of Amazonia before European conquest', *Proceedings of the Royal Society* B, vol 282, no 1812, p20150813
Cronon, W. (1996) *Uncommon ground: rethinking the human place in nature*, WW Norton & Company, New York
Denevan, W.M. (1992) 'The pristine myth: the landscape of the Americas in 1492', *Annals of the Association of American Geographers*, vol 82, no 3, pp369–385.

Ellis, E.C., Gauthier, N., Goldewijk, K.K., Bird, R.B., Boivin, N., Díaz, S., Fuller, D.Q., Gill, J.L., Kaplan, J.O., Kingston, N. and Locke, H. (2021) 'People have shaped most of terrestrial nature for at least 12,000 years' *Proceedings of the National Academy of Sciences*, vol 118, no 7, e2023483118

Funes, F., García, L., Bourque, M., Pérez, N. and Rosset, P. (2002) *Sustainable agriculture and resistance: transforming food production in Cuba*, Food First Books, Oakland, CA

Funes-Monzote, F.R., Monzote, M., Lantinga, E.A. and van Keulen, H. (2009) 'Conversion of specialised dairy farming systems into sustainable mixed farming systems in Cuba', *Environment, Development and Sustainability*, vol 11, no 4, pp765–783

Geldmann, J., Manica, A., Burgess, N.D., Coad, L. and Balmford, A. (2019) 'A global-level assessment of the effectiveness of protected areas at resisting anthropogenic pressures', *Proceedings of the National Academy of Sciences*, vol 116, no 46, pp23209–23215

Hallmann, C.A., Foppen, R.P., van Turnhout, C.A., de Kroon, H. and Jongejans, E. (2014) 'Declines in insectivorous birds are associated with high neonicotinoid concentrations', *Nature*, vol 511, no 7509, pp341–343

Hallmann, C.A., Sorg, M., Jongejans, E., Siepel, H., Hofland, N., Schwan, H., Stenmans, W., Müller, A., Sumser, H., Hörren, T. and Goulson, D. (2017) 'More than 75 percent decline over 27 years in total flying insect biomass in protected areas', *PLoS One*, vol 12, no 10, e0185809

Hardt, M. and Negri, A. (2000) *Empire*, vol. 15, Harvard University Press, Cambridge, MA

Harvey, D. (1996) *Justice, nature and the geography of difference*, Blackwell, Oxford

Iriarte, J. (2007) 'New perspectives on plant domestication and the development of agriculture in the New World', in T. Denham, J. Iriarte and L. Vrydaghs (eds) *Rethinking agriculture: archaeological and ethnoarchaeological perspectives*, Left Coast Press, Walnut, CA, pp167–188

Kolbert, E. (2014). *The sixth extinction: an unnatural history*, Henry Holt & Company, New York

Levins, R. (1969) 'Some demographic and genetic consequences of environmental heterogeneity for biological control', *American Entomologist*, vol 15, no 3, pp237–240

Levins, R. (1998) 'Rearming the revolution: the tasks of theory for hard times', *Socialism and Democracy*, vol 12, no 1, pp61–74

Lindenmayer, D.B., Wood, J., McBurney, L., Blair, D. and Banks, S.C. (2015) 'Single large versus several small: the SLOSS debate in the context of bird responses to a variable retention logging experiment', *Forest Ecology and Management*, vol 339, pp1–10

MacArthur, R.H. and Wilson, E.O. (1967) *The theory of island biogeography*, Princeton University Press, Princeton, NJ

Machín Sosa, B., Roque, A.M., Ávila, D. R. and Rosset, P.M. (2010) *Revolución agroecológica: el movimiento de campesino a campesino de la ANAP en Cuba. Cuando el Campesino ve, hace fe*, La Habana, Cuba y Jakarta, Indonesia, ANAP/La Vía Campesina/OXFAM ANAP and La Vía Campesina, La Habana

Martínez-Torres, M.E. and Rosset, P.M. (2014) 'Diálogo de saberes en La Vía Campesina: food sovereignty and agroecology', *Journal of Peasant Studies*, vol 41, no 6, pp979–997

Mather, A.S. (1992) 'The forest transition', *Area*, vol 24, pp367–379

Newmark, W.D. (1995) 'Extinction of mammal populations in western North American national parks', *Conservation Biology*, vol 9, no 3, pp512–526

O'Connor, M. (1994) *Is capitalism sustainable?: political economy and the politics of ecology*, Guilford Press, New York, NY

Perfecto, I. and Vandermeer, J. (2010) 'The agroecological matrix as alternative to the land-sparing/agriculture intensification model', *Proceedings of the National Academy of Sciences*, vol 107, no 13, pp5786–5791

Perfecto, I., Vandermeer, J. and Wright, A. (2009) *Nature's matrix: linking agriculture, conservation and food sovereignty*, 2nd Edition, Routledge, London

Phalan, B., Balmford, A., Green, R.E. and Scharlemann, J.P. (2011) 'Minimising the harm to biodiversity of producing more food globally', *Food Policy*, vol 36, ppS62–S71

van der Ploeg, J. D. (2013) *Peasants and the art of farming*, Fernwood Publishing, Halifax

van der Ploeg, J.D. (2018) *The new peasantries: rural development in times of globalization*, 2nd Edition, Routledge, London

Rosset, P.M., Machín Sosa, B., Roque Jaime, A.M. and Ávila Lozano, D.R. (2011) 'The Campesino-to-Campesino agroecology movement of ANAP in Cuba: social process methodology in the construction of sustainable peasant agriculture and food sovereignty', *The Journal of Peasant Studies*, vol 38, no 1, pp161–191

Silva, J.P., Correia, R., Alonso, H., Martins, R.C., D'Amico, M., Delgado, A., Sampaio, H., Godinho, C. and Moreira, F. (2018) 'EU protected area network did not prevent a country wide population decline in a threatened grassland bird', *PeerJ*, no 6, pe4284

Siurua, H. (2006) 'Nature above people: Rolston and "fortress" conservation in the south', *Ethics & the Environment*, vol 11, no 1, pp71–96

Toledo, V.M. (2005) 'La memoria tradicional: la importancia agroecológica de los saberes locales', *Leisa Revista de Agroecología*, vol 20, no 4, pp16–19

Tscharntke, T., Klein, A.M., Kruess, A., Steffan-Dewenter, I. and Thies, C. (2005) 'Landscape perspectives on agricultural intensification and biodiversity–ecosystem service management', *Ecology Letters*, vol 8, no 8, pp857–874

Valencia Mestre, M. C., Ferguson, B. G. and Vandermeer, J. (2018) 'Syndromes of production and tree cover dynamics of Neo-tropical grazing lands', *Agroecology and Sustainable Food Systems*, vol 43, no 4, pp362–385

Vandermeer, J. and Perfecto, I. (2012) 'Syndromes of production in agriculture: prospects for social-ecological regime change', *Ecology and Society*, vol 17, no 4, pp39–50

Vandermeer, J., Perfecto, I. and Schellhorn, N. (2010) 'Propagating sinks, ephemeral sources and percolating mosaics: conservation in landscapes', *Landscape Ecology*, vol 25, no 4, pp509–518

Wilson, E.O. (2016) *Half-earth: our planet's fight for life*, WW Norton & Company, New York

8 The background of international projects on agrobiodiversity in Cuba

Leonor Castiñeiras, Tomás Shagarodsky, Pablo Eyzaguirre and Toby Hodgkin

The traditional knowledge of farmers is derived both from their observations and their experiential learning. Today, this knowledge is vital as it assists in mitigating the crisis driven by the expansion of industrial agricultural systems which ignore the environmental, socio-cultural and economic heterogeneity of traditional farming systems. Indeed, this heterogeneity has favoured the conservation and transmission of traditional plant genetic resources from one generation to the next. Traditional agricultural systems such as home gardens and farms are able to maintain themselves over time despite environmental stresses, and this constitutes the foundation of their sustainability. This capacity depends on the characteristics of the production system, the nature and intensity of the intervening biotic and abiotic stressors and the farmers' knowledge.

Biosphere Reserves transition zones are located in proximity to the Reserves and may include human settlements, economic activity and infrastructure. They are areas where local communities, conservation institutions, scientists, nongovernmental organisations, cultural groups from different economic sectors and other stakeholders can work together towards the sustainable management of their resources. The collaboration between the Alexander von Humboldt Institute for Fundamental Research in Tropical Agriculture (INIFAT) and the International Plant Genetic Resources Institute (IPGRI, today Bioversity International) started in 1994. This collaboration has facilitated the transfer of expertise and financial assistance in the implementation of some research projects on Cuban phytogenetic resources, particularly those relating to Biosphere Reserves. Additionally, the Seville Strategy (UNESCO, 1996) – approved following the International Conference on Biosphere Reserves (1995) – created the conditions for research as it promotes valorisation of training and education in these areas, reinforcing the link between cultural and biological diversity for the conservation and sustainable use of resources.

Projects for in situ conservation of agricultural biodiversity in Cuban Biosphere Reserves

In addition to these two organisations, support was provided by other international institutions such as the Italian NGO Centro Internazionale Crocevia, the German

DOI: 10.4324/9781315183886-9

cooperation agency Deutsche Gesellschaft für Technische Zusammenarbeit (GTZ) and the Canadian agency International Development Research Centre (IDRC). Subsequent research was basically carried out within the framework of three projects (Castiñeiras et al., 1999; Watson and Eyzaguirre, 2001; Hermann et al., 2009).

- Pilot project for the *in situ* conservation of crop plant variability, led by IPGRI and Crocevia (1998).
- Contribution of home gardens to *in situ* conservation of plant genetic resources in traditional agricultural systems, led by IPGRI and GTZ (1999–2001).
- Adaptive management of seed systems and gene flow for sustainable agriculture and improved livelihood in the humid tropics of Mexico, Cuba and Peru, led by Bioversity International and IDRC (2003–2007).

The research conducted within the framework of these projects aimed to ascertain the utilisation of agricultural biodiversity (cultivated species and local cultivars) conserved by farmers for generations, and the existence of a suitable infrastructure for the development of *in situ* management and conservation programmes for these traditional plant genetic resources in line with the UNESCO's Man and the Biosphere Programme.

Pilot project for the *in situ* conservation of cultivated plant variability

This project was developed in two Cuban regions; in the west of the country, in the transition zone of the Sierra del Rosario Biosphere Reserve, and in the more central region, situated in the municipality of Cumanayagua (Cienfuegos Province) where the Cienfuegos Botanical Garden – the oldest in the country – is located. Twenty-eight home gardens and farms were involved in the project. The inventory of agricultural biodiversity in these sites listed 482 species (cultivated and wild) utilised by farmers, from 312 genera and 94 families.

The project also involved training workshops and exhibitions of traditional agricultural biodiversity held at each intervention site. This led to the establishment of a platform for the exchange of experiences between farmers, the educational sector and non-governmental organisations in the areas, such as the Cuban Association of Agricultural and Forestry Technicians (ACTAF) and the National Association of Small Farmers (ANAP). Local government representatives were also invited to join with the aim of raising awareness of the value of traditional agricultural biodiversity in existence and used by families in the home gardens and farms in these localities.

It was observed that soil preparation and weed control in the home gardens and farms is carried out using animal traction or manually. Rainfed agriculture is dominant, grown with organic fertilisation or no fertilisation at all. The best environmental health conditions – in terms of soil fertility and species management within the home garden systems – were found in the transition zone of the Sierra del Rosario Biosphere Reserve, which is an important indicator of the impact of the environmental training programme in which the farmers in that protected area participated.

Contribution of home gardens to the *in situ* conservation of plant genetic resources in traditional farming systems

Considering that Biosphere Reserves develop strategies for the conservation of natural biodiversity and for the sustainable use of resources in their transition zones, in addition to the outcomes of the previous project, the second project also included the Cuchillas del Toa Biosphere Reserve, thus covering the three geographic regions of the island (west, centre and east). Thirty-eight families from ten rural communities participated in this project, seven from the buffer zones (four in the Sierra del Rosario Biosphere Reserve and three in the Cuchillas del Toa Biosphere Reserve) and three communities from the central region (Cienfuegos province). The farms were selected after interviewing more than 140 families from different farms representing 10% of the community families. In the selection process several criteria such as the number of species cultivated on the farms (40 or more), duration of settlement in the same location (more than 20 years), family composition (more than one generation) and consent to participate in the study were adopted.

The inventory of agricultural diversity registered 508 species (about 80% cultivated and 20% wild species used by farmers), from 362 genera and 108 families. Additionally, a study of infraspecific variability of three cultivated species was conducted: *Pouteria sapota*, *Capsicum* spp. (*C. annuum*, *C. frutescens*, *C. chinense*) and *Phaseolus lunatus*. The results showed that there is a high variability, including some traditional cultivars that were previously considered extinct. These cultivars were included in the *ex situ* national collections maintained by INIFAT for conservation as a backup that can be accessed to distribute samples to the intervention sites, if necessary.

An important learning was that the Cuban home garden represents the most important element in family sustenance. Culture, climate and socio-economic status are the main factors influencing the actions and decisions of farmers to maintain a wide diversity of species and traditional management practices.

Training workshops were held by scientists with the participation of farmers from each region, and biodiversity presentations were also organised by local stakeholders. These annual meetings at each site helped support the exchange of experience on conservation practices, as well as on the management of traditional agricultural biodiversity, on the sustainability of different agricultural systems and the complementary function of *in situ* conservation of cultivated and wild species. In addition, the value of diversity in the home garden was disseminated, particularly among the political and educational authorities of the localities.

Farmers in the buffer zones of the Sierra del Rosario and Cuchillas del Toa Biosphere Reserves mainly use organic fertilisers produced from on-farm resources, such as leaves, roots, crop residues and manure from domestic animals. Planting, weeding and harvesting are primarily done manually, and both home gardens and farms are completely dependent on rainfall because they do not have access to sources of water for irrigation.

In the central region, a trend towards the substitution of traditional cultivars by commercial varieties was observed. Furthermore, farmers had a lower awareness of the infraspecific variability of crops, as they were more focused on production for the market and less on gains provided by traditional cultivars, such as different life cycles in grain cultivars providing the family with fresh produce for longer periods throughout the year. Also, a lower diversity in domestic animal species/breeds was noted, perhaps due to the location of home gardens in more anthropised areas.

Adaptive management of seed systems and gene flow for sustainable agriculture and improved livelihood in the humid tropics of Mexico, Cuba and Peru

The sites for the development of this project were the Sierra del Rosario (western region) and Cuchillas del Toa (eastern region) Biosphere Reserves. The primary objectives were to assess seed supply networks within and between communities and to identify the main constraints facing farmers from 36 farms and traditional home gardens in informal systems.

The agricultural diversity found was greater than in the previous project, despite only two sites participating in the project. A total of 555 species were recorded, from 432 genera and 240 families, which proves the higher diversity found in areas linked to Biosphere Reserves.

Traditional farming involves culture and traditions inherited from one generation to the next as well as families' income, resources available and food intake, which may vary from family to family. Socio-economic conditions combined with local knowledge, such as understanding of the most suitable soil for each crop or cultivar, or of sowing times and crop or cultivar cycles -among others-, leads to a wide range of agricultural management systems and, consequently, to the cultivation of multiple crops and great variation within the cultivated species (Hermann et al., 2009).

Four crops were selected, common beans (*Phaseolus vulgaris* L.), lima bean (*P. lunatus* L.), peppers (*Capsicum* spp.) and maize (*Zea mays* L.), to examine their infraspecific variability (Castiñeiras et al., 2006). It was observed that 90% or more of the seeds planted in family farms were sourced from the informal sector, including the farmers' own seeds. Consequently, the significance and persistence of seed supply or exchange networks in local informal systems is unquestionable.

The operation of informal seed exchange networks in rural communities is complex and depends on farmers' management and exchange skills. Seed flows can take place either into or out of the farm or home garden. In the first case, seeds come, either from another farm in the community, from neighbouring communities, or from the formal sector. In the second, seeds are delivered to another farmer. Seed exchange within the farmers' network occurs mainly within communities (about 76% of cases), while exchange between communities was about 24%, indicating that each community's seed bank is sufficient to meet the internal demands of the farmers. Seed supply and demand by farmers can vary from year to year depending on the annual harvest, which renders the informal seed system even more complex. In cases of exchange within the community, most crop seeds sown for family subsistence during the period of the study were sourced from the farmers' own stock and if required, small quantities were obtained from within the same community. The most important limitations that farmers identified in informal seed systems beyond weather conditions are economic factors such as poor marketing, transportation, or access to seed saving resources. Local and national strategies are required to improve seed quality of traditional cultivars and consolidate the farmers' production and conservation skills and contribute to the preservation of the genetic resources of traditional crops. Knowledge exchange between farmers and scientists during the training workshops – especially on seed management – led to significant increases in the participants' knowledge, primarily in improving the germination of seeds stored on the farms, with a consequential increase in crop yields.

Food production from farms and home gardens is mainly for self-consumption, although income is generated from the sale of surplus through different channels such as state entities, as is the case for coffee production in mountain areas.

Food security on farms in transition zones and buffer zones

Plant genetic resources are the biological cornerstone of food security, and their conservation is necessary to ensure agricultural production and to meet growing environmental challenges, especially climate change (FAO, 2019, 2020; Dunbar et al., 2020).

The sustainable character of home gardens and farms in transition zones makes these units economically stable; they are in balance with the environment, contain a highly diverse range of species and their traditional crops are the main sources of food for both humans and animals. The diversification and integration of food production systems into complex ecological processes create synergies with the natural habitat without depleting natural resources. Agroecology and sustainability are approaches that contribute to improving yields and increase resilience through practices such as the use of green fertilisers and sustainable soil management (FAO, 2016, 2018).

Crop plants – and their infraspecific variability – have diverse life cycles, which allows farmers to satisfy families' food and nutritional needs throughout the year and gain food security. Each crop or cultivar has a specific use (Table 8.1), and each species and its infraspecific variables are grown using different practices.

The *frijol caballero* – a traditional crop used as a legume in Cuba – is not considered a commercial crop and is cultivated only in traditional farming systems. This is a climbing species which is cultivated using just a few individuals trained over the farm fences for support. It flowers later than the common bean (*Phaseolus vulgaris* L.) and its complete reproductive cycle is also longer. Consequently, when the family has consumed the common bean harvest, they can start to consume the frijol caballero. On the other hand, *Phaseolus vulgaris* L. is a commercial crop

Table 8.1 Number of families, genera and species in the Sierra del Rosario and
Cuchillas del Toa Biosphere Reserves according to use

Use	Number of families	Number of genera	Number of species
Fruit	16	23	38
Grains	2	6	8
Roots and tubers	5	6	9
Vegetables	7	14	15
Livestock feed	8	10	10
Drinks	8	8	11
Condiments	12	21	35

which uses improved varieties. It is the main source of vegetable protein in the
Cuban diet, notwithstanding traditional varieties with excellent culinary and nutritional qualities are still cultivated in home gardens and rural farms. Other legumi-
nous plants are also cultivated on farms, such as *Vigna* spp., thus ensuring a source
of vegetable protein throughout the year.

Another example is maize, which is a good nutritional and energy source for
humans (O'Leary, 2016) and is also used – like other species – as feed for domestic animals, which ultimately contribute to satisfy the animal protein needs in the
family diet. Fruit provides the necessary vitamins and minerals (FAO, 2020), and
takes on a more important role in the family diet as it substitutes vegetables for
long periods throughout the year as these require large quantities of water during
their life cycles (in the dry season) and water is a scarce resource in traditional
farms.

Viandas such as cassava (*Manihot esculenta*), banana (*Musa* spp.), *malanga*
(*Colocasia esculenta* and *Xanthosoma* sp.), beans (*Phaseolus* spp.), maize (*Zea
mays*) and others are testimony to food customs dating back to the aboriginal cultures of Meso- and South America in which roots, tubers and grains were highly
important ingredients in the family diet. For this reason, these crops tend to occupy
larger tracts of land on the farms.

The cultivation of medicinal species is also important, and 114 species were
identified from 84 genera and 47 families for this use, as it permits families to treat
some minor health problems directly on the farm.

Development of new markets

The wealth in home gardens and farms in terms of diversity and traditional variability is vital for the food security of rural families. Most of the harvest, however,
is not commercialised – not even in local markets – which offers opportunities to
expand food availability in markets or for the development of new sales channels.

In 2006, seed and biodiversity fairs were launched to promote and sell products
and seeds representing the traditional biodiversity in the transition zones of the
Sierra del Rosario and Cuchillas del Toa Biosphere Reserves. These fairs were
held in towns close to the buffer zones with larger populations (Candelaria in the

west and Yateras in the east) to attract a larger public. Farmers had notable economic gains, although these depended on the products on offer. In addition to the economic benefit, commercialisation contributed to moderately alleviating local food deficits with products cultivated using agroecological methods. These fairs set the foundation for the commercialisation of traditional agricultural products from farms in the Biosphere Reserves, strengthening local agricultural production and promoting food sovereignty. Participating farmers were transported from east to west and vice versa (1,000 km from one site to the other) to participate in the fairs. Representatives of local government, tourism, education, agriculture and the environmental sectors, as well as the UNESCO regional and national offices in La Habana also participated in the meeting. As local governments have continued to support these marketing actions, these fairs have become a social activity that motivates farmers to offer a service to the community while receiving benefits from the diversity preserved (Shagarodsky et al., 2009).

The impact of past projects in Cuba

The interest shown by farmers and by local organisations (political, environmental, agricultural and educational) in the different processes of the study of local agricultural biodiversity in the transition zones of two Cuban Biosphere Reserves marked the start of an intersectoral collaboration with the objective of conserving phytogenetic resources. As Shandas and Messer (2008) indicate, the knowledge acquired by community members increases their commitment to conservation. Based on the results obtained, it was proposed that the identified traditional agricultural biodiversity be integrated into the conservation and management plans of the Biosphere Reserves.

Collaboration between the Cuban partners, Bioversity International and the Biosphere Reserves promoted a multidisciplinary and educational research programme aimed at the conservation of both natural and managed landscapes. Additionally, this also benefited the MaB Programme in Cuba, as it has strengthened relations between environmental experts managing the Reserves and the rural communities in the transition zones, by incorporating new elements to the Reserves' management plans and their environmental education programmes.

Furthermore, cooperation between the Cuban Ministry of Agriculture and the Biosphere Reserves has been strengthened by recognising the value of the conservation of cultivated species as part of natural biodiversity conservation strategies. Thus, Cuba's National Plant Genetic Resources Programme has benefited in three ways. First, there has been an advancement in the conservation of Cuban phytogenetic resources based on the knowledge about the management of agricultural biodiversity in the Biosphere Reserves and on the *ex situ* conservation of infraspecific variability. Second, the use and production of seeds from local agricultural biodiversity is promoted, which contributes to minimising genetic erosion. Finally, the inventory of functional both cultivated and wild species will contribute towards monitoring the conservation of traditional agricultural biodiversity. The previous advancements were taken into account in proposing *in situ* conservation strategies for Cuban agricultural biodiversity, with farmers as the main actors, within the National Action Plan on Biological Diversity of the Republic of Cuba 2006/2010 (CITMA, 2006), as well as the Strategic Plan for Biological Diversity 2011–2020 (CITMA, 2011) and the Aichi Targets (SCDB, 2011). Once the Strategic Plan was concluded – with its advances and limitations – the National Action Plan on Biological Diversity of the Republic of Cuba (CITMA, 2021) was prepared, as well as the Strategic Plan for Biological Diversity 2021–2030 (CITMA, 2021), in accordance with the global strategic vision (CITES, 2021) and the UN 2030 Agenda for Sustainable Development (2021). Cuba joined as the 66th state party to the Nagoya Protocol on Access and Benefit Sharing (ABS) in 2015.

Another important social impact was also the recognition of the farmers' role in the conservation of plant genetic resources. This led to the development of local initiatives for the conservation of cultural landscapes. This is the case of seed and traditional agricultural biodiversity fairs, with direct economic benefits for farmers (sale of seeds, agricultural products and other products derived from agrobiodiversity). Training workshops helped improve the seed quality of the crops, increasing their viability for the next planting cycle, families participated in all activities and their self-esteem increased.

The agrobiodiversity exhibitions held during the training workshops and the fairs showcasing seed and traditional agricultural products were examples of activities that encouraged exchange among farmers and an increase in diversity on the farms. Farmers took on the role of promoting the project with other rural families from neighbouring communities interested in participating. The increase in the number of species from 2001 to 2007 due to the incorporation of additional crop species through exchange with other farmers suggests a growing recognition of the benefits of diversity. This is the outcome of the work done with farmers and through environmental education aimed at the *in situ* conservation of traditional agricultural biodiversity (Castiñeiras et al., 2012). Jarvis et al. (2016) noted that the farmer is an effective agent for seed dissemination and transmission of knowledge on local diversity management. The traditional knowledge of local communities, based on their own agrobiological resources, can contribute to their economic development and increase the supply of products from traditional biodiversity to agricultural markets. Therefore, a proposal was made to extend this approach to the other four Cuban Biosphere Reserves (Guanahacabibes, Ciénaga de Zapata, Buena Vista and Baconao).

Other projects related to the management and sustainable use of agrobiodiversity in Cuban conserved landscapes

Other interventions financed with national and international funds to develop strategies linking traditional agricultural management practices with soil conservation, and with bordering habitat conservation and ecosystem management and farming in general are identified in Table 8.2.

Table 8.2 Other projects related to the management and sustainable use of agrobiodiversity in Cuban conserved landscapes

Project (funding organisation)	Description
Strengthening the National System of Protected Areas in Cuba (UNDP/GEF) (2003–2009)	A project that sought to improve the management system of the protected areas within eco-regions of global importance. The aim was to improve the strategy to achieve a higher level of integration within the National System.
Enhancing Prevention, Control and Management of Invasive Alien Species in Vulnerable Ecosystem (UNDP/GEF) (2003–2011)	The project aimed to safeguard globally important biodiversity in vulnerable ecosystems. Synergies were established with other projects integrating local agrobiodiversity on farms into environmental management plans, while promoting *in situ* conservation of local genetic resources, and reducing the impact of invasive species.
Supporting Implementation of the Cuban National Program to Combat Desertification & Drought (GEF) (2004–2015)	Aimed at capacity building with the participation of the Ministry of Agriculture (MINAG) and ANAP. It was designed to promote practices that favour the development of agrobiodiversity and contribute to mitigate soil degradation by using indigenous tree species and varieties.
Application of a Regional Approach to the Management of Marine and Coastal Protected Areas in Cuba's Southern Archipelagos (UNDP/GEF) (2005–2017)	The objective was the conservation and sustainable use of marine biodiversity of global importance through a network of marine protected areas in the archipelagos of southern Cuba. It also addressed the study of the negative impacts of agricultural practices on the coasts of the Cuchillas del Toa Biosphere Reserve, as well as the formulation of mitigation measures.
Sustainable Management of Natural Resources from the Buffer Zone of the Alexander von Humboldt National Park, Guantánamo Province (ACTAF/German Government) (2009–2013)	The main objectives were to increase economic productivity of local communities through the implementation of activities compatible with the conservation of the Alexander von Humboldt National Park, and to improve their standard of living in the buffer zone of the Cuchillas del Toa Biosphere Reserve.
Capacity Building for Planning, Decision Making and Regulatory Systems & Awareness Building/Sustainable Land Management in Severely Degraded Ecosystems (GEF) (2014–2017)	This project focused on promoting, monitoring and evaluating cross-sectoral plans to assess extreme weather events. In the framework of the project rural farmers were trained in good practices for restoring degraded land. It developed a collaborative platform for the maintenance of agrobiodiversity within Cuba's Protected Areas.

Acknowledgment

The authors wish to acknowledge to Yanisbell Sánchez due the information provided on the projects related to traditional agriculture developed in Cuba.

References

Castiñeiras, L., Barrios, O., Fernández, L., León, N., Cristóbal, R., Shagarodsky, T., Fuentes, V., Fundora, Z., Moreno, V., de Armas, D., Acuña, G., García, M., Hernández, F., Arzola, D. and Giraudy, C. (2006) *Catálogo de Cultivares Tradicionales y Nombres Locales en Fincas de las Regiones Occidental y Oriental de Cuba: Frijol Caballero, Frijol Común, Ajíes – Pimientos y Maíz*, Agrinfor, La Habana, 63pp

Castiñeiras, L., Sánchez, Y., García, M., Shagarodsky, T., Fuentes, V., Giraudy, C., Hernández, F. and Hodgkin, T. (2012) 'Oportunidades de Conservar la Biodiversidad Agrícola en las Reservas de la Biosfera de Cuba', *Naturaleza y Desarrollo*, vol 10, no 2, pp18–36

Castiñeiras, L., Shagarodsky, T. and Fundora, Z. (1999) 'Proyecto Piloto para la conservación *in situ* de la variabilidad de plantas cultivadas, Informe Final de Proyecto IPGRI/ Crocevia', Instituto de Investigaciones Fundamentales en Agricultura Tropical (INIFAT), 56pp

CITES (2021). 'Visión Estratégica de la CITES 2021-2030', Conf. 18.3 https://cites.org/ sites/default/files/document/S-Res-18-03_0.pdf, accessed November 9 2021

CITMA (2006) 'Plan de Acción Nacional sobre Diversidad Biológica de la República de Cuba 2006/2010', Ministerio de Ciencia, Tecnología y Medio Ambiente, https://www. cbd.int/doc/world/cu/cu-nbsap-v2-es.pdf, accessed November 21 2022

CITMA (2011) 'Plan Estratégico para la Diversidad Biológica de la República de Cuba 2011-2020', Ministerio de Ciencia, Tecnología y Medio Ambiente, 26pp

CITMA (2021) 'Plan Estratégico para la Diversidad Biológica de la República de Cuba 2021-2030', Ministerio de Ciencia, Tecnología y Medio Ambiente, 32pp

Dunbar, W., Subramanian, S.M., Matsumoto, I., Natori, Y., Dublin, D., Bergamini N., Mijatovic, D., González Álvarez, A., Yiu, E., Ichikawa, K., Morimoto, Y., Halewood, M., Maundu, P., Salvemini, D., Tschenscher, T. and Mock, G. (2020) 'Lessons learned from application of the indicators of resilience in socio-ecological production landscapes and seascapes (SEPLS)', in O. Saito, S.M. Subramanian, S. Hashimoto and K. Takeuchi (eds), *Managing socio-ecological production landscapes and seascapes for sustainable communities in Asia. Mapping and navigating stakeholders, policy and action*, Science for Sustainable Societies, https://doi.org/10.1007/978-981-15-1133-2_6, acccessed September 11 2022

FAO (2016) *El Estado Mundial de la Agricultura y la Alimentación. Cambio Climático, Agricultura y Seguridad Alimentaria*, http://www.fao.org/publications/sofa/2016/es/, accessed January 10 2022

FAO (2018) *Panorama de la Seguridad Alimentaria y Nutricional – Desigualdad y Sistemas Alimentarios*, www.fao.org/3/CA2127ES/ca2127es.pdf, accessed November–October 10 2022

FAO (2019) *Comisión de Recursos Genéticos para la Alimentación y la Agricultura*, https:// www.fao.org/cgrfa/topics/plants/es, accessed May 5 2022

FAO (2020) *Frutas y verduras - esenciales en tu dieta. Año Internacional de las Frutas y Verduras 2021*, Documento de antecedentes, Rome, https://www.fao.org/3/cb2395es/ CB2395ES.pdf, accessed November 8 2021

Hermann, M., Amaya, K., Latournerie, L. and Castiñeiras, L. (2009) ¿Cómo conservan los agricultores sus semillas en el trópico húmedo de Cuba, México y Perú? Experiencias de un Proyecto de Investigación en Sistemas Informales de Semillas de Chile, Frijoles y Maíz, Bioversity International, Rome, 179pp

Jarvis, D.I., Hodgkin, T., Brown, A.H.D., Tuxil, J., Lopez, I., Smale, M. and Sthapit, B. (2016), *Crop genetic diversity in the field and on the farm: principles and applications in research practices*, Yale University Press, 416pp

O'Leary, M. (2016) *Maíz: De México para el mundo*, https://www.cimmyt.org/es/uncategorized/maiz-de-mexico-para-el-mundo/, accessed November 10 2021

SCDB (2011) 'Plan Estratégico para la Diversidad Biológica 2011-2020 y las Metas de Aichi', http://www.cbd.int/doc/strategic-plan/2011-2020/Aichi-Targets-ES.pdf, accessed March 12 2022

Shagarodsky, T., Arias, L., Castiñeiras, L., García, M. and Giraudy, C. (2009) 'Ferias de agrobiodiversidad y semillas como apoyo a la conservación de la biodiversidad en Cuba y México', in M. Hermann, K. Amaya, L. Latournerie and L. Castiñeiras (eds) *¿Cómo conservan los agricultores sus semillas en el trópico húmedo de Cuba, México y Perú? Experiencias de un Proyecto de Investigación en Sistemas Informales de Semillas de Chile, Frijoles y Maíz*, Bioversity International, Rome, pp102–122

Shandas, V. and Messer, W.B. (2008) 'Fostering green communities though civic engagement', *Journal of the American Planning Association*, vol 74, no 4, pp408–418

UNESCO (1996) *Reservas de la Biosfera: La Estrategia de Sevilla y el Marco Estatutario de la Red Mundial*, UNESCO, Paris

Watson, J.W. and Eyzaguirre, P.B. (2001) *Proceedings of the second international home gardens*, Workshop, 17–19 July 2001, Witzenhausen, Federal Republic of Germany

9 Measuring farm environmental sustainability in the Sierra del Rosario Biosphere Reserve

Gaia Gullotta, Nadia Bergamini, Paola De Santis, Alejandro González Álvarez, Jorge Luis Zamora Martín, José Manuel Guzmán Menéndez, Nicola Tormen and Enrico Ruzzier

Introduction

Following the goal of all-round sustainability, biodiversity (both cultivated and wild) is returning as a central element of farming. Biodiversity is a fundamental element of the farm as a complex, and contributes, through ecosystem services, to the plasticity and resilience in case of adverse events. In this context, Man and Biosphere (MaB) Reserves represent a good example of integration between human communities and the biosphere, and the MaB reserve of Sierra del Rosario, given the vicissitudes that have characterized the Cuban economy in the last 40 years, is an example of integration between people, agriculture and nature. Cuban traditional farming system is the result of the abandonment of production practices associated with both the economic boom and the green revolution, in favour of traditional and empirical practices, only apparently less efficient. On the contrary, they are fully functional for small-scale production, and adapted to a scarcity of means and resources. In the Sierra del Rosario agriculture is located to varying degrees on the edge and within the reserve. As farmers were driven, on the one hand, by the need to survive in spite of the hardships that followed the collapse of the communist regimes in Eastern Europe and the US *embargo*, and on the other by the restrictions in the reserve itself. However, despite the awareness of the scientific community of this reality, no complete characterization or evaluation of these systems has been performed (Castiñeiras et al., 2001).

Family farming, sustainability and landscape connectivity: a new approach

An integrated and low cost methodology was applied to characterize farms in terms of sustainable agriculture (FAO, 2022), their contribution to biodiversity and agro-biodiversity conservation, and to landscape structural connectivity (Tischendorf and Fahrig, 2000). Building on The Biodiversity Friend protocol (Caoduro et al., 2014), originally envisaged both to assess European agriculture systems and to certificate products from sustainable agriculture, the method was further developed

DOI: 10.4324/9781315183886-10

and adapted to family farming systems in the tropics, by including cultural and socio-economic variables as well as attributes of farm structure and natural elements. Data on biodiversity, both wild species and cultivated varieties, agricultural practices, environmental quality, and on the social, economic and cultural dimensions were collected through diverse methods, including a household semi-structured survey, the calculation of soil biodiversity index and vegetation *relevés*. Thus, the information provided by farmers through the survey was proven by direct observations and measurements. Technical experts from the Instituto de Investigaciones Fundamentales en Agricultura Tropical (INIFAT), Sierra del Rosario Biosphere Reserve (SRBR), Ministerio de Ciencia, Tecnología y Medio Ambiente (CITMA), World Biodiversity Association (WBA) and Bioversity International organized and carried out the fieldwork with the active participation of the farmers who, in addition to being the subjects of the survey, were also involved in data collection. Collection of plant and soil samples was performed both during the dry and wet seasons (in March and July respectively) to reveal possible differences in the composition of plant and animal communities associated to seasonality. In order to gain robust evidence, data collection was carried out in various farms, both inside and outside of the Biosphere Reserve. Variables measuring farmer management practices, soil biodiversity and number and type of plant species identified were weighted and scored to determine the impact of agriculture on biodiversity and three different levels of sustainability are identified (low, medium and high). Good management practices, as here understood, include activities with minimal impact on wild animal and plant species in the agroecosystem.

Selection of the most representative farms of Sierra del Rosario

Twenty-four farms in SRBR and its surrounding area (as control group) were selected for this research study on the basis of four criteria: contribution to structural landscape connectivity, type of landscape unit, main farming activity, and distance from the Reserve's centroid. According to the first criterion, farms were classified as per high, medium and low connectivity based on their distance to seven forest patterns applying the Morphological Spatial Pattern Analysis (MSPA) (Vogt et al., 2007) and using a 2015 Landsat 8 image (Landsat 8-Landsat Missions, 2017) with a spatial resolution of 30 m. The second criterion was applied on the geomorphological, physical and biological characteristics of the landscape units, including dominant vegetation, as the type of landscape unit influences farm structure and farm management choices (Figure 9.1). The study area has four landscape units: two highland areas and two flatlands, described in Table 9.1.

Farms with different levels of connectivity (first criterion) were selected from each landscape unit. The third criterion guided the selection of the farms based on the main farming activity, which allowed for a better understanding of the role of farming systems in the conservation of biodiversity and landscape connectivity. In Sierra del Rosario coffee plantations and animal husbandry are the dominant farming activities in the highlands, while lowlands are used for mixed crops and livestock. The distance from the centroid of the protected area was used as the

Table 9.1 Description of the landscape units identified

Unit	
Highland 1	Horst anticlines with slopes between 0° and 45°, strongly dissected formed by limestone on shallow and stony reddish brown ferrosols soils, with a predominance of semi-deciduous forest, evergreen, secondary vegetation.
Highland 2	Denudative and erosive litho-structural heights, with slopes from 0° to 25°, strongly dissected formed by limestone, flysch and metamorphic rocks on shallow and stony leached reddish-brown ferrosols soils with a predominance of secondary semi-deciduous and evergreen forest and secondary vegetation.
Flatland 1	Fluvio-marine plain, slightly inclined (0°–10°), ondulated, strongly dissected formed by limestone and marl on shallow and stony dark-reddish ferrosol soils with a predominance of grasses, royal palm, crops and secondary vegetation, with some patches of semi-deciduous forest and secondary gallery vegetation.
Flatland 2	Erosive-denudative plain, slightly inclined (0°–10°), highly dissected, hilly, formed by limestone, sandstone, flysch and volcanic rock, on shallow and stony brown soils with dark-reddish ferrosols carbonates and alluvial deposits, with a predominance of secondary vegetation and royal palms, and pastures.

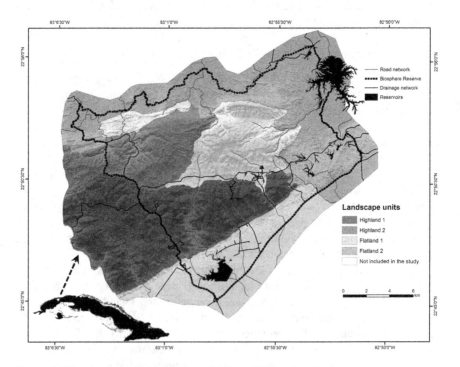

Figure 9.1 Landscape units in the Sierra del Rosario Biosphere Reserve.

fourth criterion, and both the furthest and the closest farms to the centroid were selected from each landscape unit. This allowed to compare the structural connectivity among farms with similar characteristics but geographically distant. Following such criteria, six farms in each landscape unit were selected, amounting to a total of 24 farms.

Field survey

A household semi-structured survey with 48 questions, clustered in 10 sections (Table 9.2), was developed expanding the Biodiversity Friend protocol with additional questions on cultural and socio-economic variables, farm structure and natural elements, and applied to all 24 selected farmers.

General information about the farmer (e.g., name, sex and age) and the farm (e.g., cooperative name, status of usufruct, farm area and geographic coordinates) was collected under the first section.

Questions in sections 2 through 4 provided information on constraints (e.g., pests and diseases) and management practices adopted by the farmers. Pest and disease control ranged from the use of integrated management strategies to synthetic pesticides, crop rotation and manual removal of pest organisms. The most harmful pests identified are *broca del café* (*Hypothenemus hampei*), which mainly affects coffee plantations, and *bibijagua* (*Atta insularis*) mainly affecting cassava (*Manihot esculenta*) and citrus fruit (*Citrus* sp.). On de other hand, *dormidera* (*Mimosa pudica*) and *guisazo* (*Xanthium strumarium*) are the most harmful plants which threaten crops and pastures decreasing quantity and quality of forage (Ajorlo et al., 2014; Saeed et al., 2020). Farmers adopt low impact practices to mitigate biotic stress, including the cultivation of *flor de muerto* (*Tagetes erecta*) to control nematodes, or the usage of barriers and live fences to prevent pests from entering crop fields since they serve as refuge for beneficial insects. The infected plants are removed and mostly used as combustible, for green manure or dead barriers when dried.

Questions on land uses and farm management revealed that forestry and agroforestry patches prevail in the highlands of Sierra del Rosario, while the flatlands

Table 9.2 Household survey sections

Section
General information
Pest and disease management
Land use and farm management
Soil management
Water management
Elements with high natural value
Agricultural and structural biodiversity
Cultural aspects
Livestock management
Socio-economic aspects

are mainly covered by pastures. The shade coffee plantation is a mixed agroforestry system, characterized by a multiple strata structure: herbs, shrubs and trees. Therefore, coffee is grown under a canopy of various species such as fruit trees, timber trees and medicinal plants, combining native and introduced species (Moguel and Toledo, 1999). Another common land patch in the farms is the home garden (*huerto casero*) which plays an important role for biodiversity conservation and as reservoirs of traditional knowledge (Eyzaguirre and Linares, 2004). The home garden hosts a high diversity of medicinal plants, spices, herbs and ornamental plants despite its relatively small extent (Castiñeiras et al., 2001). In the SRBR, women often take care of the home garden, becoming also the repositories of traditional knowledge about the use, cultivation and conservation of the plant species grown there. Despite the practicality of a register of the farm operations, it is not a common practice among the farmers interviewed. Farmers also informed on waste management. Most of them recycle plastic materials and glass, but the rest of the waste is burned or left in the field.

Regarding soil management, leaves of Cuban royal palm (*Roystonea regia*) are used as a barrier to protect soil from erosion. Other examples of good soil management practices were provided by a farmer, who grows *cucaracha* (*Tradescantia zebrina*) as cover crop. The same farmer also informed about the use of river sediments where the white ginger lily (*Hedychium coronarium*) grows, rich in nutrients and suitable for seed germination and plant propagation. This is an exotic invasive species, but if properly managed through periodical cuts, has beneficial effect on the soil. Concerning fertilizers, all types are used, inorganic, organic derived from animal or vegetable matter, or synthetic organic such as urea.

Water management adopts practices associated to a rain fed agricultural system in a karst landscape in a context of climate change, where scenarios predict drier conditions for Central America and the Caribbean in the future (FAO, 2016). Most farmers perceive a general increase in temperature and decrease in precipitation, and a consequence is that crops, like rice, that require great amounts of water are no longer cultivated in Sierra del Rosario. The most common source of water are creeks (*cañadas*) with a semi-permanent water regime. Some farmers harvest rainwater, but only a few farmers have large artificial ponds.

The survey also investigated the elements that increase landscape connectivity, matrix complexity and biodiversity – such as composition and change in number and size over time of live fences. In Sierra del Rosario, the most common plants used as live fences are *almácigo* (*Bursera simaruba*) and *piñón florido* (*Gliricidia sepium*). Some live fences have two layers, shrubs and trees. Thus, some farmers planted *almácigo* together with *piña* (*Ananas comosus*) in order to have a multifunctional live fence that operates both as barrier for livestock and for the production of fruit. Another type of ecological corridor is shaped by small groups of trees. Farmers grow mainly fruit trees as shade trees and for fruit production within the cultivated land patches. Dry-stone walls also have a high natural value. They maintain a micro-climate suitable for certain plant and animal communities that increase the biodiversity and act as corridors connecting habitats. Farmers usually use dry-stone walls to prevent soil erosion and to build riverbanks. Finally, to complete the section on landscape connectivity, the role of pollinators was evaluated. Wild and artificial beehives with the two allochthonous species *Melipona* (*Melipona beecheii*) and *Española* (*Apis mellifera iberiensis*) are common in the area. *M. beecheii* is a threatened species in the Continental America due to the habitat loss and the competition with the introduced African bees. In contrast, in Cuba *M. beecheii* population is actually increasing as most of the farmers maintain one or more beehives of this species in their home gardens, and use bee products for self-consumption, especially for medicinal purposes.

Farmers also informed on the use of woodlands (secondary forests). In addition to obtaining firewood and charcoal, they are a source of wood for construction, furniture, fences and tools. The forest also provides medicinal plants, organic matter to produce fertilizer, and animal feed, as livestock often has free access for grazing. The level of grazing affects plant growth, community structure, ecosystems services and functioning and can impact natural forest regeneration and biodiversity in general (Díaz et al, 2007). Most of the farmers breed livestock both for home-consumption and for sale, especially cattle, pigs and poultry. Only few farmers practice intensive livestock farming. Additionally, some farmers hunt wild animals such as pigeons and Jutía conga (*Capromys pilorides*). The *hutia* is captured and fed for human consumption. Some farmers hunt not only mammals and birds but also fish and reptiles such as Cuban cichlid (*Nandopsis tetracanthus*) and *Cuban maja* (*Chilabothrus angulifer*) respectively.

More than half of the farmers interviewed had additional sources of income such as with the sale of charcoal, pensions from previous employment, contracts with the government for forest management and off-farm sources of income by other family members, such as education, tourism and forest management.

The rich agricultural and forest biodiversity (species, varieties), and their products, is the result of a long process of selection by culturally diverse farmers and forest communities across the world. Among women's and men's differentiated priorities, access to, and benefits from agricultural and forest biodiversity influence their resources management. Results show that women at the SRBR are basically involved in all farm activities, with the exception of heaviest physical labour. Yet, home garden management and seed conservation and selection are almost exclusively a female job.

Farming and soil biodiversity conservation

The impact of agriculture on soil biodiversity was evaluated using the Soil Biodiversity Index (IBS-bf) (Caoduro et al., 2014). This method assesses the level of conservation of the soil-related invertebrate community through the detection of specific bioindicators, called umbrella morphogroups (Caoduro et al., 2014; Menta et al., 2015). In the index, umbrella morphogroups are ranked and weighed based on their importance as bioindicators. What is examined in the survey is the mere presence of a morphogroup, not the relative abundance.

Two samples were collected in each farm, as a combination of 3 smaller random subsamples. The soil collected, once sifted with a solid reducer in order to separate the fine debris from the coarser ones, was then distributed on a white sheet for direct observation. Morphogroups were then noted on the record module for index calculation. Following the Biodiversity Friend method, a specific score/weight (20, 15, 10, 5) is assigned to each morphogroup, depending on the importance of that specific bioindicator (important groups score higher) (Caoduro et al., 2014).

The farms with higher IBS-bf values have proven to be those with permanent land cover, reduced use of synthetic fertilizers, mixed vegetation associations with the presence of coffee; while, on the other hand, the farms with the lowest values corresponded to those with less land cover and a high livestock load.

Plant diversity on-farm

Wild and cultivated plants from the different land patches such as pasture, secondary forest, coffee plantation, fallow land, were studied through vegetation *relevés*. The Braun-Blanquet cover-abundance scale (Braun-Blanquet, 1932) was adopted to evaluate the relative abundance of each species. Additional data such as geographic coordinates, soil texture, vegetation structure (the percentage of coverage of woods, shrubs and grasslands and their average height) and the main perturbations (e.g. grazing, slash and burn) were collected for each plot. This method allowed not only to assess the plant diversity on-farm, but also to characterize the land uses, and compare the Sierra del Rosario Biosphere Reserve with its surrounding area. More than 300 different species were identified out of over 2,000 plants sampled across the 24 selected farms. Out of the 300 species 224 are endemic, autochthonous and naturalized. The highest richness and relative abundance of species were found in riparian vegetation, coffee plantations and fruit orchards.

Among all the farms surveyed only few adopt a sub-intensive agricultural system. These farms are all located outside the boundaries of the reserve. Farms within the reserve have a high degree of naturality and contribute substantially to the conservation of biodiversity, agrobiodiversity and landscape connectivity. Of particular relevance is the contribution of the farms growing coffee in the highlands of the reserve. With a semi-wild cultivation system, shade coffee plantations are well integrated in the natural environment and their multi-strata vegetation substantially contribute to landscape structural connectivity. In addition, the canopy of the coffee helps to protect the soil from the intense solar radiation. Combined with the accumulation of leaves and other organic matter, the shade coffee plantation system favours the maintenance of a humid and deep soil litter, and hosts a complex community of soil invertebrates.

Conclusion

The multidisciplinary approach adopted has proved to be effective to evaluate the sustainability of farming activities, their contribution to landscape connectivity and the impacts of Cuban traditional farms on biodiversity. This methodology has the advantages of being low cost (simple and easy to use equipment) and requires minimum training. The scoring system of the evaluation helps the farmers identify the strengths and weakness of their management practices and provides guidance to adopt and enhance the practices that favour sustainability and biodiversity conservation.

The involvement of the farmers in the survey and data collection contributed to improving their understanding of conservation and its links with sustainability. This approach placed farmers at the heart of the system, highlighting their key role in biodiversity and agricultural landscape conservation.

This methodology combines observations and measurements from different research areas that allow a comprehensive assessment to analyse the degree of compatibility of farming systems with conservation approaches. The preliminary results show a high degree of naturality and sustainability, particularly in the shade coffee plantations in the highlands that combine traditional varieties and knowledge and host rich biodiversity.

Acknowledgments

We would like to thank the participating communities of the Asentamiento el Establo, Bahía Honda, Candito Frías, Carambola, Cayajabos, Coublet, El Carmen, La Comadre, La Lechuza, Mango Bonito Mantilla, San Francisco, San Miguel Coblet, Soroa, Valdés and the staff of the Estación Ecológica Sierra del Rosario.

References

Ajorlo, M., Abdullah, R., Abdul Halim, R. and Ebrahimian, M. (2014) 'Cattle grazing effect on *Mimosa pudica* L. in tropical pasture system', *Pertanika Journal of Tropical Agricultural Science*, vol 37, no 2, pp249–261

Braun-Blanquet, J. (1932) *Plant sociology*, McGraw-Hill Book Company, New York

Caoduro, G., Battiston, R., Giachino, P.M., Guidolin, L. and Lazzarin, G. (2014) 'Biodiversity indices for the assessment of air, water and soil quality of the "Biodiversity Friend" certification in temperate areas', *Biodiversity Journal*, vol 5, no 1, pp69–86

Castiñeiras, L., Fundora Mayor, Z. and Shagarodsky, T. (2001) 'Contribution of home gardens to *in situ* conservation of plant genetic resources in farming systems – Cuban component', in J.W. Watson and P. Eyzaguirre (eds) *Home gardens and in situ conservation of plant genetic resources in farming systems*, International Plant Genetic Resources Institute (IPGRI), Rome, pp42–56

Díaz, S., Lavorel, S., McIntyre, S., Falczuk, V., Casanoves, F., Milchunas, D.G., Skarpe, C., Rusch, G., Sternberg, M., Noy-Meir, I., Landsberg, J., Zhang, W., Clark, H. and Campbell, B.D. (2007) 'Plant trait responses to grazing - a global synthesis', *Global Change Biology*, vol 13, no 2, pp313–341, doi: 10.1111/j.1365-2486.2006.01288.x, accessed October 20 2022

Eyzaguirre, P.B. and Linares, O.F. (2004) *Home gardens and agrobiodiversity*, Smithsonian, Washington, DC

FAO (2016) *Drought characteristics and management in the Caribbean*, FAO Water Reports 42, FAO, Rome

FAO (2022) 'Sustainable food and agriculture', https://www.fao.org/sustainability/en/, accessed November 4 2022

Landsat 8 - Landsat Missions (2017), https://landsat.usgs.gov/landsat-8, accessed October 27 2017

Menta, C., Tagliapietra, A., Caoduro, G., Zanetti, A. and Pinto, S. (2015) 'Ibs-Bf and Qbs-Ar comparison: two quantitative indices based on soil fauna community', *EC Agriculture,* vol 2, no 5, pp427–439

Moguel, P. and Toledo, V.M. (1999) 'Biodiversity conservation in traditional coffee systems of Mexico', *Conservation Biology*, vol 13, no 1, pp11–22, doi: 10.1046/j.1523-1739.1999.97153.x, accessed November 4 2022

Saeed, A., Hussain, A., Khan, M.I., Arif, M., Maqbool, M.M., Mehmood, H., Iqbal, M., Alkahtani, J. and Elshikh, M.S. (2020) 'The influence of environmental factors on seed germination of *Xanthium strumarium* L.: implications for management', *PLoS One*, vol 15, no 10, pe0241601, doi: 10.1371/journal.pone.0241601

Tischendorf, L. and Fahrig, L. (2000) 'On the usage and measurement of landscape connectivity', *Oikos*, vol 90, no 1, pp7–19, doi: 10.1034/j.1600-0706.2000.900102.x

Vogt, P., Riitters, K.H., Estreguil, C., Kozak, J., Wade, T.G. and Wickham, J.D. (2007) 'Mapping spatial patterns with morphological image processing', *Landscape Ecology*, vol 22, no 2, pp171–177, doi: 10.1007/s10980-006-9013-2

10 Ecosystem services in agrobiodiversity and family farms

Indicators of social and ecological resilience

Alejandro González Álvarez, Nadia Bergamini, Dunja Mijatovic and Yanisbell Sánchez Rodríguez

Monoculture agroecosystems generally have a lower capacity to respond to environmental perturbations such as drought, floods, pest outbreaks or the presence of invasive species, as well as to other uncertainties deriving from market fluctuations. Analogously, diversified agroecosystems that lose values due to its conversion and intensification of land use are also cause for concern regarding their functioning and ability to respond adaptively to external shocks (Ceroni et al., 2011).

Traditional family farming constitutes a productive type of agroecosystem capable of providing both food and means of production. In the case of Latin America, Salcedo and Guzmán (2011) consider family farming as that practised on "a farm of sufficient size to provide the sustenance of a family and where paid labour is not required as the farm is worked by members of one's own family". Traditional family farming is not only important for its productive value, but also for its symbolic nature. People frequently come into contact with biodiversity on farms which are often the main pool of cultural traditions and local know-how (Jarvis et al., 2011).

The ecological and symbolic wealth of traditional family farming resides in its capacity to model a range of diverse landscapes. In every corner of the world, farming areas have been chiselled out through interaction between people and nature over time (Brown et al., 2005). These landscapes integrate the cultural and natural heritage of the people who live there and often have global value for the conservation of biodiversity (Lino and Britto de Morales, 2005). Protecting these landscapes requires a conservation approach that acknowledges their natural and sociocultural values, while fostering traditional connections with the land and involving local people in its stewardship.

The concept of resilience is used in the field of systems ecology as an explanatory approach for adaptation processes when faced with perturbations. The most recent applications of this concept incorporate the role of human societies in the transformation of ecosystems. In this way, the concept of social-ecological systems is used to illustrate this interdependence and to examine how different societies establish more or less resilient resource management, generating an adaptive and organisational management model capable of responding to changes. Social

DOI: 10.4324/9781315183886-11

networks and collective memory are recognised as important sources of resilience, constituting the basis on which to develop knowledge-based creativity (Fernández and Morán, 2012).

There is a wide range in the diversity and resilience that characterise the heterogeneity of social-ecological systems present in the world. At one extreme, industrial agriculture – dependent on high inputs and focused on the profitable generation of goods – constitutes the antithesis of social-ecological resilience. This is attributable to the inherent weakness of these systems regarding the generation and maintenance of ecosystem services, since these do not have any market value (Ceroni et al., 2011). Such services (understood as the functions of the ecosystems from which human beings benefit) are organised into four wide groups: provisioning services (food, water, wood and fibres); regulating services (climate and spread of disease); cultural services (contributing to the spiritual well-being of people) and supporting services (soil formation, quality of air or water) (FAO, 2021).

At the other extreme, traditional family farming offers a resilient alternative, both from a social and environmental point of view. The high degree of independence that this system can achieve in a scenario of financial capital restraints implies that the family farmer bases the generation of wealth on his/her own labour, and therefore depends only on services made available by nature (van der Ploeg, 2013). Thus, family farming is seen as a multifunctional and more sustainable form of agriculture, maintaining ecosystem functions and processes such as soil erosion control, carbon sequestration, nutrient recycling, habitat generation for wildlife and sources for spiritual and cultural recreation (Ceroni et al., 2011). As follows, society is the principal beneficiary of these services, which frequently take the form of productive landscapes in biosphere reserves where predominantly inhabited areas coexist and integrate with other predominantly natural areas. Notwithstanding, measuring resilience at a landscape scale implies enormous methodological challenges, since the resilience of a complex system is an emerging property conditioned by the specific interaction between farmers, the farm, and its context (Cabell and Oelofse, 2012).

Application of social-ecological indicators in Cuba

An indicator is a significant physical, chemical, biological, social, or economic variable that can be measured for operational purposes (Brown and Hodgkin, 2011). Thus, as part of the Satoyama initiative, indicators of social-ecological resilience were applied in the Sierra del Rosario and Cuchillas del Toa biosphere reserves. This initiative (inspired by a Japanese term used to name traditional rural landscapes) is multi-institutional and backed by the Japanese Ministry of the Environment. A remarkable aspect of the Satoyama indicators is that they have not been conceived exclusively as a measurement tool, with the challenges that this implies, but essentially as a vehicle for understanding and strengthening the resilience of a site. Cuba has been a pioneer country in the application of such social-ecological indicators in the Cuchillas del Toa Biosphere Reserve. The initiative examines several groups of indicators (Table 10.1): (1) protection of ecosystems and maintenance of diversity; (2) agricultural biodiversity; (3) knowledge, learning and innovation; and (4) social equity and infrastructure (Bergamini et al., 2013). To measure each aspect within the group of indicators, farmers were asked to attribute a value from 1 to 5, as well as to report on the perceived trend over time: increasing, decreasing or stable. In Cuchillas del Toa, twelve people from four different communities were interviewed, five landowners (four men and one woman), in addition to two non-owner women and five youngsters. In the Sierra del Rosario Biosphere Reserve, the indicators of the Satoyama initiative were also evaluated (Table 10.1), although there was a difference in the methodology as a group session approach was adopted instead of individual interviews. The sample consisted of ten people, all adult rural farmers, including two women.

When evaluating resilience in social-ecological landscapes it is important to identify vulnerabilities as a first step towards reducing them (Cabell and Oelofse, 2012). In the case of the Cuchillas del Toa Biosphere Reserve, one of the aspects identified for consolidating resilience is the improvement of infrastructures associated with services, particularly in activities such as energy generation, transportation and access to telecommunications, which were identified as important limitations for the permanence of people in the communities, especially young people. Among the vulnerabilities identified after applying the indicators in Sierra del Rosario was the need to improve the management of both liquid and solid waste, as well as access to services such as transportation. Some changes took place following the sharing of results of the workshop organised with decision makers, whereas others require more complex intervention. In general, the rating

Table 10.1 Resilience indicators of the Satoyama initiative.

Group of indicators		Examined issues
1	Protection of ecosystems and conservation of biodiversity	Heterogeneity and multifunctionality of the landscape, sustainable use of resources, protected areas due to their ecological and cultural importance, environmental safety and health; Recovery capacity against external impacts
2	Agricultural biodiversity	Maintenance, documentation and conservation of biodiversity in the community; Diversity of the local food ecosystem and local consumption
3	Knowledge, learning and innovation	Innovation in the management of agricultural biodiversity to improve resilience and sustainability; Access and exchange of agricultural biodiversity; Transmission of traditional knowledge from elders, parents and relatives to new generations in the community; Cultural traditions related to biodiversity; Number of generations interacting with the landscape; Documentation practices and exchange of traditional knowledge; Use of local terms or indigenous languages; Knowledge of women regarding biodiversity and its uses
4	Social equity and infrastructure	Management of local resources; Rights/Autonomy in relation to land and resource management; Gender; Social infrastructure; Access to health care; Health risks, Means of support and income, Mobility

obtained by the evaluated social-ecological landscapes was high, presenting either a stable or primarily favourable trend. In particular, the results for Sierra del Rosario – compared with other countries of the global south (Bolivia, Iran, Sri Lanka, Zimbabwe, India) where similar surveys were conducted – demonstrate that the Cuban biosphere reserve obtains higher scores, as well as more favourable trends in relation to the other social-ecological landscapes.

An effect of the application of these indicators is also academic. This practice leads to new paradigms that help understand the complexity of agricultural systems and go beyond the essentially agronomic paradigm of the conservation of plant genetic resources, formerly prevalent in previous works of the Instituto de Investigaciones Fundamentales en Agricultura Tropical Alejandro de Humboldt (INIFAT). Undoubtedly, understanding and strengthening resilience requires a multifaceted perspective: ecological, socioeconomic, and plant genetics. These indicators offer an opportunity for further analysis and study.

Lessons learned and results in the study sites

The strengths identified must be attributed largely to the sense of belonging expressed by the participants in relation to the place where they live and is consubstantial with their continuity in the agricultural landscape with which they interact. This also comes from the social work that the country has developed and the growing interest of the general public in agroecology and agrobiodiversity.

The application of the first group of indicators (Table 10.1), corroborates that many of the agroecosystems of the two biosphere reserves are integrated within a high-quality matrix (Figure 10.1) in accordance with the diverse use of land, alternating between agroforestry areas, crop fields and grazing areas. A relevant issue is the value of diversification, considered a support factor in the adaptive capacity of agricultural systems from both an ecological and economic perspective.

Within the context of a family agricultural unit integrated into a social-ecological landscape, it is feasible to strengthen resilience through the recognition and promotion of species that have a fundamental or support role for the generation of income and opportunities for the well-being of the farmer. This is related in turn to the structural diversity of an agroecosystem. In this sense, in many cases the agroecosystems in Sierra del Rosario and Cuchillas del Toa have the capacity to export agricultural goods that contribute to the food security of society as a whole. Among the products that generate surpluses in family farming in the two reserves, we can identify coffee (*Coffea* spp.), malanga (*Xanthosoma* spp.), coconut (*Cocos nucifera* L.) and beans (*Phaseolus vulgaris* L.). However, this capacity does not compromise the value of use of most of the traditional crops, thus contributing to a high level of food self-sufficiency.

On the other hand, diversification is closely related to the capacity of the social-ecological system to recover from extreme events, such as hurricanes. Despite reports of recent impacts in the two biosphere reserves, at the time of the application of the Sierra del Rosario indicators, the most recent previous impacts were from hurricanes Gustave and Ike, which landed in the reserves for a short interval

Figure 10.1 Social-ecological productive landscape with different matrices.

in 2008. The effects on domestic and natural vegetation were significant, as they were on fauna and the landscape. Despite this, the loss of human lives was minimal thanks to the system of Civil Protection which warned and evacuated the population. Many people hosted relatives and friends in their own homes. When the hurricanes passed, reconstruction started immediately with the intervention of the Civil Protection and other national institutions as well as other local civil society organisations (CSOs) that had already played a very important role in response. Local knowledge was also vital such as in the case of the traditional type of construction in the region known as a *vara en tierra* maintained by some rural inhabitants and which guarantees a safe place in the event of a hurricane.

However, not everything is negative following the impact of a hurricane, according to some farmers who also reported positive aspects. For example, the 2008 hurricanes led to a re-establishment in the levels of the water table, which facilitated the cultivation of vegetables and the production of new crops. The fallen trees became a source to produce charcoal or organic matter. The exchange of seeds between farmers helped strengthen social ties and the re-establishment of local agricultural systems. State support and international aid (roofs, filters, mattresses) also contributed to the social-economic reactivation of the community.

The evaluation of agrobiodiversity in both reserves using the second group of indicators (Table 10.1), confirmed that the resilience of the communities could be strengthened by increasing plant genetic diversity. Thus, a functional perspective – both ecological and economic – in the strategies to be adopted must be encouraged.

Agroecology emphasises the diversification of agroecosystems as a strategy for building resilience, prioritising those irreplaceable species for the ecological functioning of an agroecosystem: legumes, pollinators or pest controls, which are essentially linked to the provision of essential ecosystem services or regulation (Córdova-Tapia and Zambrano, 2015). The disappearance of these key species greatly impacts fundamental ecological processes. Recognising that it is not possible to radically modify the structural diversity of farms in Cuba due to their social purpose, the strategy adopted is to enrich the variety of plant species which is highly significant for functional diversity. At the same time, due to their high unit value, they are an important factor in the potential generation of income with respect to economically consolidated agricultural activities. An example of this is the introduction or promotion of species such as: *Azadirachta indica* A. Juss (Neem), a source of biopesticides for pest control, *Juglans jamaicensis* C. DC (West Indian walnut), a native species with an ecological function in natural ecosystems, as well as for timber, and *Cedrela odorata* L. (Cigar box cedar). Other examples to point out are that of *Cinnamomum cassia* (L.) J. Presl (*Canela china*), a tree with a high unit value, or *Tamarindus indica* L. (Tamarind), fruit and legume species with nitrogen-enriching properties for the soil.

An important question of agricultural diversity in family units is its variability between social-ecological contexts, and therefore it is important to differentiate such context and its comparative advantages and existing types of production. The promotion of a resilient agrobiodiversity must consider which plant genetic resources can be appropriate for assimilation in a given context. It is also important to take into account for each landscape, the presence of urban spaces, especially useful in promoting the utilisation of vegetables or of underutilised species that have comparative market access advantages.

Within the third group of indicators, the farmers reported on the use of a group of traditional farming practices. In particular, the distribution of crops on hillsides according to their orientation, so crops that are more tolerant to water stress such as maize (*Zea mays* L.) and pineapple (*Ananas comosus* (L.) Merr.), are planted on south-facing slopes, more exposed to the sun; while crops that are more dependent on humidity such as malanga (*Xanthosoma* spp.), coffee (*Coffea arabica* L.) and French plantain (*Musa paradisiaca L.*) are planted on north-facing slopes, that receive more shade (Figure 10.2).

In terms of seed conservation, there are some traditional practices and knowledge that are more readily open to scientific discussion. With this in mind, consolidating the technological capacities of rural farmers is one of many actions taken to consolidate resilience at a local level. Through workshops and training courses, technological knowledge has been transmitted to strengthen pest management, the conservation of local seeds, or the improvement of soil fertility. It has been reported, however, that there is great potential for innovation in farming families in areas such as incorporating new genetic resources or management techniques or participating in activities such as agrotourism or cottage industries.

The fourth group of indicators leads us to note that promoting social resilience of rural populations is a very important strategy for safeguarding the agrobiodiversity within certain landscapes. Assuming the premise that the farmer family unit is a key structure in the configuration and conservation of agrobiodiversity, it is essential that this unit also be safeguarded. In the family farming economy, in which labour is even more fundamental than financial capital (van der Ploeg, 2013), a population with high educational and health standards is the most efficient way

Figure 10.2 Banana cultivation (*Musa paradisiaca* L.) on a north-facing slope.

of increasing resilience that a developing country can provide for its agricultural sector. A farming sector with high educational standards has a greater capacity for learning, innovation and adaptation, self-organisation and self-sufficiency, which is also considered part of the generation of resilience. (Fernández and Morán, 2012). Similarly, in the generation of social-ecological resilience, development actions such as the electrification of rural communities in biosphere reserves is necessary. Although this does not contribute directly to the increase in product diversification in these areas, indirectly it is a fundamental contribution since when this service is introduced in a remote community – as seen from experience – it helps create a stable local population that will act on safeguarding local agrobiodiversity.

The duality of family farming systems as generators of goods and services is an argument in defence of their protection, especially in the face of other options that southern countries face for their development, such as mining. Although this last activity guarantees a significant return on capital, it irreversibly compromises the generation of services and goods in a landscape in the short and medium term. In Cuba, there are examples of environmental debates that have been won in favour of the preservation of water sources against mining, which demonstrates the viability of sustainable development.

Finally, it is important to increase the self-esteem of the family farmer within landscapes from a heritage perspective, through international programmes that offer opportunities to acknowledge and safeguard the cultural, environmental and technological value of agrarian landscapes with social-ecological significance (Koohafkan and Altieri, 2016). Linking Cuba with the *Important World Agricultural Heritage Systems* programme, provides an alternative for the adoption of strategies for the generation of added value in the context of social-ecological landscapes in protected areas of managed resources.

Acknowledgments

The authors wish to acknowledge the contribution of the Bolivian specialist Helga Gruberg in the application of social-ecological indicators in Sierra del Rosario, as well as the staff of her Ecological Station. Similarly, they are also grateful to the specialists from INIFAT and the Provincial Unit of Environmental Services of Guantánamo who participated in the project.

References

Bergamini, N., Blasiak, R., Eyzaguirre, P., Ichikawa, K., Mijatovic, D., Nakao, F. and Subramanian, S.M. (2013) *Indicators of resilience in socio-ecological production landscapes*, United Nations University Institute of Advanced Studies, Tokyo

Brown, A.H. and Hodgkin, T. (2011) 'Medición, manejo y mantenimiento en fincas de la diversidad genética de los cultivos. Manejo de la biodiversidad en los sistemas agrícolas', in D. Jarvis, C. Padoch and H.D. Cooper (eds) *La biodiversidad, la agricultura y los servicios ambientales. Manejo de la biodiversidad en los sistemas agrícolas*, Bioversity International, Rome, pp14–36

Brown, J., Mitchell, N. and Baresford, M. (2005) *The protected landscape approach: linking nature, culture and community*, Gland, IUCN, 268pp

Cabell, J.F. and Oelofse, M. (2012) 'An indicator framework for assessing agroecosystem resilience', *Ecology and Society*, vol 17, no 1, p18

Ceroni, M., Liu, S. and Costanza, R. (2011) 'Papeles ecológico y económico de la biodiversidad en los agroecosistemas. Manejo de la biodiversidad en los sistemas agrícolas', in D. Jarvis, C. Padoch and H.D. Cooper (eds) *La biodiversidad, la agricultura y los servicios ambientales. Manejo de la biodiversidad en los sistemas agrícolas*, Bioversity International, Rome, pp475–503

Córdova-Tapia, F. and Zambrano, L. (2015) 'La diversidad funcional en la ecología de comunidades', *Ecosistemas*, vol 24, no 3, pp78–87

FAO (2021) 'Servicios Ecosistémicos y Biodiversidad', FAO, Rome, http://www.fao.org/ecosystem-services-biodiversity. Consultado en 17/12/2021, accessed July 12 2022

Fernández, J.L. and Morán, N. (2012) 'Cultivar la resiliencia. Los aportes de la agricultura urbana a las ciudades en transición', *Papeles de relaciones ecosociales y cambio global*, no 119, pp131–143

Jarvis, D., Padoch, C. and Cooper, H.D. (2011) *La biodiversidad, la agricultura y los servicios ambientales. Manejo de la biodiversidad en los sistemas agrícolas*, Bioversity International, Rome

Koohafkan, P. and Altieri, M. (2016) *Forgotten agricultural heritage. Reconnecting food systems and sustainable development*, Routledge, 296pp

Lino, C.F. and Britto de Morales, M. (2005) 'Protecting landscapes and seascapes: experience from coastal regions of Brazil', in J. Brown, N. Mitchell and M. Baresford (eds) *The protected landscape approach: linking nature, culture and community*, Gland, IUCN, pp163–176

Salcedo, S. and Guzmán, L. (2014) *Agricultura familiar en América Latina y el Caribe. Recomendaciones de políticas*, FAO, Santiago de Chile, http://www.fao.org/fileadmin/user_upload/AGRO_Noticias/docsRecomendacionesPolAgrFAMLAC.pdf, accessed December 17 2021

van der Ploeg, J.D. (2013) *Peasants and the art of farming. A Chayanovian manifiesto*, Fernwood Publishing, Canada, 144pp

11 Family farms as innovative living labs in agroecology and transfer of knowledge

Madeleine Kaufmann, Alejandro González Álvarez, Alberto Tarraza and Alessandra Giuliani

Introduction

The recent shift towards the inclusion of human interests in nature conservation signals a broadening perspective among conservationists. Earlier practices of fencing off pieces of nature to mitigate human influence proved to be unsustainable regarding social and conservation impacts (Lui et al., 2001; Adams et al., 2004). One of the first and best-known concepts aiming to reconcile biodiversity conservation with its sustainable use is the UNESCO's Man and the Biosphere reserve Programme (MaB). In the UNESCO MaB reserves around the world, some of the richest and diverse agricultural landscapes can be found. Yet, while biodiversity conservation has always been a fundamental principle in the management of the MaB reserves, the protection of agricultural landscapes with its agricultural biodiversity, has largely been neglected (Lenné and Wood, 2011). Agrobiodiversity is not only important because agricultural systems do heavily affect the conservation of wild biodiversity, but it has also been proved that farming landscapes host and maintain in different ways a large share of the planet's biodiversity (Gemmill, 2001). Landscapes rich in agrobiodiversity are often the product of complex farming systems that have developed in response to the unique physical conditions of a given location, such as altitude, slopes, soils, climates, combined with farming communities' cultural and social dimensions (Amend et al., 2008).

An important example is this regard is the Cuchillas del Toa MaB Reserve, located in Cuba's most eastern Province Guantánamo, where smallholder farmers manage the coexistence of natural vegetation with on-farm cash and subsistence crops and combine managed and natural landscapes in a synergistic way (Sánchez et al., 2012). The traditional varieties and their wild relatives are cultivated in the Cuban small-scale agricultural system, locally called *conuco* (Castiñeiras et al., 2002). Farmers in their *conucos* have maintained to a certain extent traditional varieties even after the introduction of modern cultivars and the promotion of monocultures of imported varieties by centralized, state-run development projects (UNEP 2010). The family farmers living in the core, buffer and transition zones of the Cuchillas del Toa MaB reserve practice traditional eco-agriculture and home garden cultivation based on their extensive knowledge of agricultural resources

DOI: 10.4324/9781315183886-12

and the sustainable use of natural vegetation (Castiñeiras et al., 2002; Sánchez et al., 2012). Today, these farming systems host some varieties that are extremely valuable as they can be exclusively found in the Cuchillas del Toa MaB reserve, e.g. landraces of beans (*Phaseolus vulgaris*) and maize (*Zea mays*) (UNEP-WCMC, 2010).

However, farmers get little benefits from managing and maintaining on-farm agrobiodiversity, resulting in little incentives to contribute actively to the sustainable functioning of the Cuchillas del Toa MaB reserve. This being said, it becomes evident that the practical reality of the Man and the Biosphere reserves' conceptual idea to implement dual "conservation" and "sustainable use and development" goals is challenging (Coetzer et al., 2014).

This case study aims to contribute to a better understanding of the many different values of the Cuchillas del Toa MaB reserve area, including the services it can provide to realize the full potential of a landscape rich in agrobiodiversity, and to investigate the agricultural landscapes in the Cuchillas del Toa MaB reserve, their extent, location, management practices, species composition and interaction, in two different agroecological zones (mountain and coastal) in the Cuchillas del Toa MaB reserve in Cuba. The prevailing farming systems were assessed, the agricultural practices described and the socio-economic factors that influence farmers' livelihood with regard to agrobiodiversity conservation identified.

The research was done in the framework of the project Agrobiodiversity Conservation and Man and the Biosphere Reserves in Cuba: Bridging Managed and Natural Landscapes (COBARB), funded by the United Nations Environment Program (UNEP) and the Global Environment Facility (GEF) Trust Fund, and managed by Bioversity International and the Instituto de Investigaciones Fundamentales en Agricultura Tropical Alejandro de Humboldt (INIFAT). Over seven months in 2013/2014 37 farmers in two distinctive agroecological zones (coastal and mountain area) of the Cuchillas del Toa MaB reserve were visited and interviewed. They live in 15 rural communities in the reserve's core, buffer and transition zones. While the communities in the coastal agroecological zone are located within the core zone (Alejandro de Humboldt National Park), those in the mountain agroecological zone are either located in the buffer or the transition zones of the reserve. Table 11.1 gives an overview of the visited communities and their location in the agroecological (mountain and coastal area) and MaB reserve zone (core, buffer and transition zone).

Various fieldwork methods were combined, including a quantitative household-level survey, key informant interviews, focus group discussions and a qualitative survey on wild diversity. A purposive, non-random sampling was applied. The research ended with a participatory farmer workshop to identify opportunities, synergies and trade-offs considering the use of agrobiodiversity as an option to improve the conservation of protected areas. Since the present research is more descriptive and explanatory in nature, descriptive statistics were obtained.

Table 11.1 Farming communities and their location in the agroecological and MaB reserve zone of the Cuchillas del Toa MaB reserve

Community	Municipality	Agroecological zone	MaB reserve zone
La Bamba	Yateras	Mountain	Transition zone
La Demahagua	Yateras	Mountain	Transition zone
La Fangosa	Manuel Tames	Mountain	Transition zone
La Jaiba	Yateras	Mountain	Transition zone
Las Municiones	Yateras	Mountain	Buffer zone
La Vuelta	Manuel Tames	Mountain	Transition zone
Rancho de Yagua	Manuel Tames	Mountain	Transition zone
Rincones	Yateras	Mountain	Transition zone
Saburén	Yateras	Mountain	Transition zone
Vega Grande	Manuel Tames	Mountain	Transition zone
Miramar	Baracoa	Coastal	Core zone
Nibujíon	Baracoa	Coastal	Core zone
Nuevo Mundo	Baracoa	Coastal	Core zone
Recreo	Baracoa	Coastal	Core zone
Santa María	Baracoa	Coastal	Core zone

Attributes of agroecological farming systems

As critical differences between various types of agricultural systems in terms of their potential to support agrobiodiversity conservation exist, the diversity of farming systems was found to be an important indicator when it comes to assess the quality of the Cuchillas del Toa agroecological zones and farming systems. To proceed with the analysis, two agroecological zones were defined: a mountain and a coastal agroecological zone (Table 11.2).[1]

Based on the interviews with 37 households, it has been found that farmers maintain both traditional varieties and their wild relatives, manage the diversity through their use, and select them according to the necessities of their families. These varieties are grown in different traditional mixed farming systems. Farmers in the mountain agroecological zone combine arable farming with the raising of livestock (mainly chicken and pigs, local landraces). Plants are cultivated in plots of land next to the farmhouse, followed by farming in forests with secondary vegetation (Figure 11.1a). Varieties *negro* and *colorado* of common bean (*Phaseolus vulgaris*), maize (*Zea mays)*, Yellow Guinea yam (*Dioscorea cayenensis*), tomato (*Solanum lycopersicum*) and cabbage (*Brassica oleracea*) are the most important cash crops. Households diversify their income by producing animal meat sold to the cooperative or intermediaries. Local landraces of chicken and pigs also serve as an important protein source for farming households. Another predominant farming system is shade-grown coffee. The composition of coffee plantations is determined by the regional coffee enterprise, with a density of around 4,000 coffee plants (*Coffea arabica*) and 250 fruit trees, e.g. avocado (*Persea americana*), sweet lime (*Citrus limetta*), sweet orange (*Citrus sinensis*), mandarin (*Citrus reticulata*) and banana (*Musa* spp.), per hectare. The coffee enterprise also determines the shade vegetation

Table 11.2 Summary of the agro-climatic characteristics of the two agroecological zones (coastal and mountain) in the Cuchillas del Toa MaB reserve

Characteristics	Coastal agroecological zone (I)	Mountain agroecological zone (II)
MaB area	Core zone	Buffer and transition zone
Altitude (m a.s.l.)	0–50	400–800
Average temperature (°C)	24–26	21–24
Relative humidity (%)	81–83	83–87
Annual rainfall (mm)	1,400–2,800	1,300–2,100
Slope (degrees)	0–25	5–60
Vegetation types	Mangroves, coastal shrubland, mesophyll evergreen forest of low altitude, submontane sclerophyllous rain forest on serpentine, rain forest of low altitude, and small patches of forest and secondary shrubland with degraded plantations	Sub-mountain mesophyll evergreen forest, pine plantations, pine forest of *Pinus cubensis*, xerófilo de Mogote, mesophyll evergreen forest of low altitude, and small patches secondary forest and shrubland with degraded plantations
Soil types	Cambisols, ferralsols and luvisols	Cambisols, ferralsols, luvisols and leptosols
Accessibility (hours)[a]	2.3	5.2

[a] Accessibility was estimated as travel time to the next bigger city (Guantanamo).

Figure 11.1 (a) Example of landscape in the mountain agroecological zone. Farmers predominately grow grains (maize and beans), root and tuber crops, and a high diversity of fruit tree species and raise livestock (sheep, goats, pigs and chicken). (b) Example of coconut plantation in the farming community of Nuevo Mundo, located in the coastal agroecological zone. Coconut trees are grown very densely with fruit trees and bushes as understory.

Source: M. Kaufmann.

species, such as mountain immortelle (*Erythrina poeppigiana*), Cigar box cedar (*Cedrela odorata*) and Jamaican nettle tree (*Trema micrantha*).

   The prevailing farming system in the coastal-agroecological zone is the coconut plantation. Large areas covered by coconut tree (*Cocos nucifera*) are found along the coast while smaller stands inter-mixed with secondary vegetation occur on hilly terrain (Figure 11.1b). Climatic conditions favour fruit production and farmers commercialize citrus fruits and pineapples through their cooperatives. Livestock (chicken, pigs, sheep and goats) is mainly raised for household consumption; however, some of the farmers seek to upscale their production to have a second income pillar, beside coconut production.

On-farm species and varieties and wild flora management

All traditional farming systems, out of their management intensity, contribute to safeguarding natural habitats and to the conservation of wild species, which leads to a high degree of plant diversity. This is consistent with research results by Amend et al. (2008) and Lockie and Carpenter (2010). Both observed that landscapes rich in agrobiodiversity are often the product of mixed farming systems that have developed in response to the unique physical conditions of a given location. Farming families use a wide range of wild flora species for their multiple benefits, exemplifying the farmers' contribution to maintain and protect wild diversity in the reserve. Farmers in both zones use wild flora for several purposes, e.g. as living fences with *pajúa* palm (*Bactris cubensis*), shade in coffee plantations with mountain immortelle (*Erythrina poeppigiana*), pest control in home gardens with French marigold (*Tagetes patula*), or fruit production collected by female farmers in patches with natural vegetation, as in the case of pineapple (*Ananas comosus*) escaped from cultivation and sapote (*Pouteria sapota*) that are important vitamin sources for households' nutrition. Several forest trees are used by farmers for dwelling construction and commercial use, e.g. Cuban royal palm (*Roystonea regia*), Cigar box cedar (*Cedrela odorata*) and *majagua* (*Talipariti elatum*), or for firewood, e.g. cupania (*Cupania juglandifolia*), copey (*Clusia rosea*) and sickle bush (*Dichrostachys cinerea*). Farmers also grow species that cattle, sheep and goats can graze, e.g. Napier grass (*Cenchrus purpureus*), *jaraguá* grass (*Hyparrhenia rufa*). Farmers with coffee and coconut plantations identified wild flora species that contribute to maintain soil humidity in their plots, e.g. silver inch plant (*Tradescantia zebrina*), and *cortadera* (*Scleria* sp.).

In 37% of the farms, patches of evergreen forest – multipurpose patches – are intercropped with different fruit trees and plants, e.g. mango (*Mangifera indica*), avocado (*Persea americana*), coffee (*Coffea* spp.) and Cuban royal palms (*Roystonea regia*), among others (Figure 11.2). These multipurpose patches are good examples of how farmers manage the coexistence of natural vegetation with on-farm cash and subsistence crops. Several flora species are used for multiple purposes, e.g. bitter orange (*Citrus aurantium*) as fresh fruit for human consumption, as medicinal tea against cold or for pest control. This reflects the sustainable use of agrobiodiversity, as the chance for substitution of these species is lower, compared to others with a single use. They are not only important examples that illustrate the high degree of on-farm agrobiodiversity but also serve as important elements to improve the quality of the agricultural and landscape matrix. Evidence for such multipurpose use of species can be found in the literature. Jambulingam and Fernandes (1986) found guava (*Psidium guajava*) and tamarind (*Tamarindus indica*)

Figure 11.2 Components of the multipurpose agricultural and landscape matrix (zone II) in
the Cuchillas del Toa Man and Biosphere Reserve.

Credit: M. Kaufmann.

were used as timber, firewood, fodder for livestock and for fruit production by
farmers in India. Another example is the African star-chestnut (*Sterculia africana*)
and the white star apple (*Gambeya albida*) grown by smallholder farmers in Nige-
ria for their fruit and used as medicinal plant and for timber (Faith Aladi and Olujobi Olaguju, 2014).

In these multipurpose patches farmers also cultivate a variety of neglected and
underutilized species (NUS), among other reasons, to diversify their diet, e.g. fruit
species (*Annona reticulata, Annona squamosa, Chrysophyllum cainito*), food col-
ouring species (*Bixa orellana*) or condiments, such as *Cinnamomum cassia, Capsicum chinense,* and *Capsicum frutescens*. Farmers in the most remote mountain
agroecological zone cultivate the highest NUS diversity (13.9 NUS species on
average per farm). The poor accessibility, and the cultural traits brought by past
Haitian migrants are thinkable causes. In other zones, NUS as stigmatized as "food
of the poor", a label observed in other regions of the world (Giuliani, 2007; Hermann et al., 2013; Padulosi et al., 2013).

Traditional knowledge and on-farm agrobiodiversity

Family farmers were found to have extensive knowledge of local plants and their
environments. It was observed that they have been using a variety of indigenous

fauna species to fulfil the socio-cultural and spiritual needs of their families.[2] This so-called traditional knowledge, i.e. the knowledge, know-how and practices that are developed, sustained and passed on from generation to generation within a community (Laurens, 2021), of family farms in the Cuchillas del Toa MaB reserve can be seen as key factor in shaping plant diversity and plays an important role in agrobiodiversity management. As in other areas of the world (FAO, 2004; Sing and Sureja, 2006), conservation and use of local agrobiodiversity was linked to different cultural traits and food habits. Several farmers cultivate plants with magic-religious purposes like snakeroot (*Eupatorium* sp.), or *rompebatallas* (*Koanophyllon villosum*), purported to provide farm prosperity. The Franco-Haitian ancestry of the farming families introduced a number of Haitian plants, still maintained as part of their diet (*Abelmochus esculentus* by 82% and *Luffa acutangula* by 19% of these farming families). Farming women collect diverse indigenous fruits in forest land, shifting lands and common lands. The *güira* (*Crescentia cujete*) fruit, for instance, is a common example, of which the beverage *miel de güira* is prepared with the

Figure 11.3 The beverage *miel de güira* (see bottle), an example of practice of farmers' traditional knowledge. Note: This beverage is said to have purifying properties. Dried güira fruit skins are used as bowls to store food.

Source: M. Kaufmann.

pulp of the fruit. It is of widespread use, with 57% of farming households using it regularly (Figure 11.3). As a component of ethnomedicine, it is said to have purifying properties and is often used by women as fertility treatment.

Local knowledge is embedded in social structures, based on experience, adapted to the environment and continuously developing (FAO, 2004). Therefore, the research aimed to identify important influencing factors for traditional knowledge, as well as other socio-economic driving forces influencing farmer's livelihood and thus the diversity of cultivated and wild species by family farmers in the Cuchillas del Toa MaB reserve, being crucial to the aimed better understanding of the many different kinds of values of the Cuchillas del Toa MaB reserve area can provide.

Results show a generally lower level of agrobiodiversity on farms that are more commercially oriented, which confirm Anderson's (2003) findings, Traditional subsistence farming systems have more diversity of inputs, resources and outputs than commercially-oriented farms and therefore maintain a higher diversity of species and varieties. Marketing channels, closely related to the degree of remoteness, play an important role too. According to farmers, the most remote mountain agroecological zone has a poor transportation infrastructure and accessibility, limiting options to sell their farm products, and suitable market channels only exist for produce that are part of the Cuban basic food basket (e.g. common beans). Smallholder farmers consequently focus their production mainly on few cash crops that secure their sale. This contributes to a reduction of the on-farm agrobiodiversity. Apart from the orientation of the farm (commercial vs. subsistence), age is another important social factor that shapes local traditional knowledge. Older people were found to be highly important repositories of traditional knowledge, particularly regarding medicinal plants and species cultivated for their spiritual and magic-religious purposes. Generally, family farms maintained by old farmers were found to have a higher level of agrobiodiversity.

Furthermore, rural flight, driven by limited livelihood options (e.g. access to education and medical care), raises concerns about intergenerational transfer of tenure over farm holding (see also Lobley et al., 2010) and to loss of agrobiodiversity in sections of the study area. As farmers do no longer see farming continuity, they focus their production on fewer species and varieties, most commonly on those market-oriented through the cooperative. As women are often the key repositories of traditional knowledge for sustainable resource management (FAO, 2004), a highly pronounced female out-migration (in most cases with their kids for better access to education and health services) in the most remote located mountain-agroecological zone is taking place. While literature generally refers to male out-migration to urban areas, leaving the women behind taking over traditional men's roles (McEvoy, 2008; Wooten 2003; Maharjan et al., 2012), in the observed communities, the situation is just the opposite. According to interviewed single-men households, female out-migration has already led to agrobiodiversity loss on their farms, as they simplify their food habits and limit their diet to two bean varieties, one species of root and tuber crop and rice. Ornamental plants, different fruit trees or species domesticated for special purposes are no longer cultivated. *Sagú* (*Maranta arundinacea*), for instance, produced for its starch obtained from the rootstock and given to babies and small children as fresh milk substitute has widely disappeared (found only on 5.4% of the interviewed farms). Farmers observed that the diversity of cultivated and wild species in their small family farms has decreased over the last decades.

Conclusions

Many species in the Cuchillas del Toa MaB reserve are only maintained by family farming, highlighting their important contribution to maintain and protect agrobiodiversity in the reserve. Family farms thus play a pivotal role in the conservation of natural and managed biodiversity, as well as in providing diversified food and food security in general.

Aside of this contribution, family farmers face a number of challenges and threats when it comes to the preservation of on-farm agrobiodiversity. The most important is the trade-off between maintaining the traditional farming system and

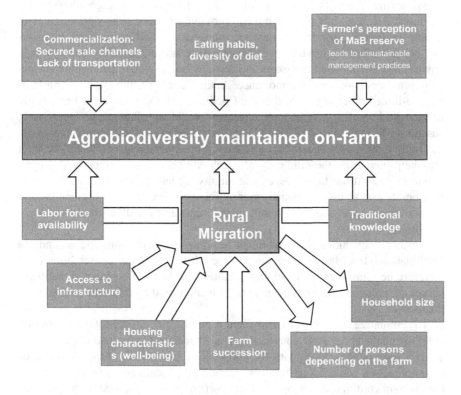

Figure 11.4 Socio-economic factors intervening in on-farm agrobiodiversity. Some of them have a direct influence (e.g. eating habits and commercialization), others influence it indirectly by fostering rural migration (e.g. access to basic infrastructure). Rural migration also negatively affects the level of farmers' traditional knowledge and the availability of on-farm labour force.

sustaining their livelihood. External pressures and limited services undermine the ability of family farmers to sustain the system and agrobiodiversity, fostering migration and uncertain intergenerational transfer of tenure. Lack of work force, poor transportation infrastructure limits the sale of farm products, and promote the concentration on very few cash crops. This also may lead to a rapid extinction of farmers' traditional knowledge (Figure 11.4).

Despite the inadequate livelihood options and a strong "urban bias" with seemingly better economic and living opportunities in the cities, it turned out that interviewed farmers have high emotional attachments to their homes, families and communities and repeatedly expressed their wish to stay in the area. Interviews carried out with farmers in zones I and II showed that only little improvements in basic infrastructure would be needed to motivate them to stay on their farms or to migrate back from urban areas to the countryside (e.g. a newly established mini-hydroelectric plant in zone II, through a private initiative of a farmer, guaranteed the access to electricity for the majority of farmers in the zone, and has motivated four farming families to migrate back from Guantánamo city so far.

Mitigating those factors that lower agrobiodiversity conservation in the Cuchillas del Toa MaB reserve implies providing farmers with opportunities to obtain increased gains from landscape management and agricultural biodiversity conservation. Improving road infrastructure and accessibility and invest in environmental education programs and capacity building for sustainable management practices is of special importance. At the same time, the creation of a MaB reserve-linked label to ensure that the Cuchillas del Toa values and benefits are recognized by policy makers and consumers would contribute as an incentive to sustain on-farm agrobiodiversity. Given the close link between agrobiodiversity and traditional knowledge, the preservation, protection and promotion of the latter may help increasing the efficiency of agrobiodiversity management and conservation efforts. At the farmer level, community leadership and local initiatives to collaborate and share experiences more effectively, create new partnerships and integrate the conservation of cultural and spiritual values at all levels are necessary conditions.

It can be concluded that traditional farming systems within and around the Cuchillas del Toa MaB reserve, compatible with biodiversity conservation must therefore be reinvigorated, its multiple values assessed and be ensured that these values are recognized by policy makers, in markets and by the rural communities themselves.

To summarize it can be said, that there is no one-size-fits-all solution to implement the challenging dual "conservation" and "sustainable use and development" goals a reserve area. Only with the adoption of innovative, adaptive and flexible approaches, while taking into consideration the local needs and conditions of the farming communities in the different parts of the reserve sustainable agricultural practices and biodiversity may continue. All legally important aspects must be defined to ensure the environmental protection and farmers at the same time provided with valuable opportunities to obtain benefits from agrobiodiversity conservation.

Notes

1 According to FAO (1996), agroecological zoning refers to the division of an area of land into smaller units with similar characteristics. Agroecological zones are defined as areas "which have similar combinations of climate, landforms, soil characteristics and land cover, as well as similar physical potentials for agricultural production [...] and similar combination of constraints and potentials for land use and development (Ibid.)".
2 The recording of traditional knowledge was based on direct observation and is therefore qualitative.

References

Adams, W.M., Aveling, R., Brockington, D., Dickson, B., Elliot, J., Hutton, J., Roe, D., Vira B. and Wolmer, W. (2004) 'Biodiversity conservation and the eradication of poverty', *Science Bulletin of the Faculty of Agriculture*, vol 306, no 5699, pp1146–1149

Amend, T., Brown, J., Kothari, A., Phillips, A. and Stolton, S. (2008) *Protected landscapes and agrobiodiversity values*, IUCN and GTZ, Kasparek Verlag, Heidelberg, 144pp

Anderson, S. (2003) 'Sustaining Livelihoods through Genetic Resources Conservation', in FAO (ed) *Training Manual on Building on gender, agrobiodiversity and local knowledge*, Food and Agriculture Organization, Rome, pp1–5

Castiñeiras, L., Fundora Mayor, Z., Shagarodsky, T., Moreno, V., Barrios, O., Fernandez, L. and Cristobal, R. (2002) 'Contribution of home gardens to *in situ* conservation of plant genetic resources in farming systems – Cuban component', in J.W. Watson and P. Eyzaguirre (eds) *Home gardens and in situ conservation of plant genetic resources in farming systems*, Smithsonian Books, Washington, DC, pp42–56

Coetzer, K.L., Witkowski, E.T.F. and Barend, F.N. (2014) 'Reviewing Biosphere Reserves globally: effective conservation action or bureaucratic label?' *Biological Reviews*, vol 89, no 1, pp82–104

Faith Aladi, S. and Olujobi Olaguju, J. (2014) 'Farmers' perception of opportunities preferences and obstacles of growing multipurpose trees on farmland in Kogi state', *European Scientific Journal*, vol 10, no 14, pp1–11

FAO (1996) *Agro-ecological zoning: guidelines*, FAO soils Bulletin 73, Food and Agriculture Organization, Rome, 93pp

FAO (2004) *Building on gender, agrobiodiversity and local knowledge*, Food and Agriculture Organization, Rome, 4pp

Gemmill, B.(comp) (2001) *Managing agricultural resources for biodiversity conservation: A guide to best practices*, UNEP/UNDP Biodiversity Planning Support Programe, Nairobi, 63pp

Giuliani, A. (2007) *Developing markets for agrobiodiversity: securing livelihoods in dryland areas*, Earthscan, London

Hermann, M., Kwek, M.J., Khoo, T.K. and Amaya, K. (2013) 'Collective action towards enhanced knowledge management of neglected and underutilized species: making use of internet opportunities', *Acta Horticulturae*, vol 979, no 1, pp65–77

Jambulingam, R. and Fernandes, E.C.M. (1986) 'Multipurpose trees and shrubs in Tamil Nadu State (India)', *Agroforestry Systems*, vol 4, no 1, pp17–32

Laurens, J.P. (2021) *Reading life with Gwich'in: an educational approach*, Routledge, London

Lenné, J.M. and Wood, D. (2011) *Agrobiodiversity management for food security: a critical review*, CAB International, Cambridge, 239pp

Lobley, M., Baker, J.R. and Whitehead, I. (2010) 'Farm succession and retirement: some international comparisons', *Journal of Agriculture, Food Systems and Community Development*, vol 1, no 1, pp49–64

Lockie, S. and Carpenter, D. (2010) *Agriculture, biodiversity and markets: livelihoods and agroecology in comparative perspective*, Earthscan, London, 318pp

Lui, J., Linderman, M., Ouyang, Z., An, L., Yang, J. and Zhang, H. (2001) 'Ecological degradation in protected areas: the case of Wolong nature reserve for giant pandas', *Science Bulletin of the Faculty of Agriculture*, vol 292, no 5514, pp98–101

Maharjan, A., Bauer, S. and Knerr, B. (2012) 'Do rural women who stay behind benefit from male out-migration: a case study in the hills of Nepal', *Gender, Technology and Development*, vol 16, no 1, pp95–123

McEvoy, J.P. (2008) 'Male Out-Migration and the Women Left Behind: A Case Study of a Small Farming Community in Southeastern Mexico', Master thesis, Utah State University, http://digitalcommons.usu.edu/etd/179, accessed June 9 2022

Padulosi, S. Bala Ravi, S., Rojas, W., Valdivia, R., Jager, M., Polar, V., Gotor, E. and Bhag, M. (2013) 'Experiences and lessons learned in the framework of a global UN effort in support of neglected and underutilized species', *Acta Horticulturae*, vol 979, no 1, pp517–532

Sánchez, Y., Castiñeiras, L., Barrios, O., González, A., González Chávez, M., de Armas, D., Socorro, A. and Cristóbal, R. (2012) 'Socio-ecological production landscapes in Cuchillas del Toa Biosphere Reserve', http://satoyama-initiative.org/en/socio-ecological-production-landscapes-in-cuchillas-del-toa-biosphere-reserve/, accessed March 20 2020

Singh, R.K. and Sureja, A.K. (2006) 'Community knowledge and sustainable natural resources management: Learning from Monpa tribe of Arunachal Pradesh', *The Journal for Transdisciplinary Research in South Africa*, vol 2, no 1, pp73–102

UNEP (2010) *Agrobiodiversity conservation and man and the biosphere reserves in Cuba: bridging managed and natural landscapes*, United Nations Environment Programme, Nairobi, 160pp

UNEP-WCMC (2010) *A-Z guide of areas of biodiversity importance*, United Nations Environment Programme-World Conservation Monitoring Centre, http://www.biodiversitya-z.org/, accessed March 15 2022

Wooten, S. (2003) 'Women, men and market gardens: gender relations and income generation in rural mali', *Rural Organization,* vol 62, no 2, pp166–177

12 The *conuco* in Cuban agricultural systems and its contribution to agrobiodiversity

Alejandro González Álvarez, Parviz Koohafkhan and Tomás Shagarodsky

Agriculture in Cuba dates back more than two thousand years when native peoples from different regions of the American continent identified plant resources and developed ecosystem management practices that gave rise to the *conuco*, a farm for the subsistence of the family and with a high level of biodiversity that already supplied the better part of the livelihood of the early inhabitants of the island (Pérez Cruz, 2014). Although not a primary centre of plant domestication, the island is a centre of secondary diversity for various domesticated species (MINAG, 2007), and of functional agricultural systems with profound historical roots and relevance to the formation of landscapes, which are a living agricultural heritage. These agricultural systems and landscapes have been shaped and maintained by generations of farmers, herders and fisherpeople around the world through interaction with diverse ecosystems using creative management practices and techniques. Thus, recognizing the importance of these agricultural systems and the contribution of family farmers and indigenous communities to global food supply and food security is a major task for their protection and revitalization (Koohafkhan and Altieri, 2016).

Pre-Columbian agriculture

At the time of the exploration of America by southern Europeans in 1492, the largest Caribbean island did not host any civilization comparable politically and technologically to those of Mesoamerica or the Andes. Nevertheless, the aboriginal populations of the island – notably culturally heterogeneous – had remarkable agricultural skills as well as a great capacity in the management and use of biological resources (Pérez Cruz, 2014). In proto-agricultural communities the existence of a mixed economy has been documented, which included the cultivation of sweet potatoes (*Ipomoea batatas*), *malanga* (*Xanthosoma* sp.; *Colocasia* sp.), Yellow Guinea yam (*Dioscorea* spp.), maize (*Zea mays*), peanut (*Arachis hypogaea*) and different species of legumes (beans) as significant sources of food (Carratalá, 2014). Among the agricultural populations the Arhuaco group (Lucayos, Ciboneyes, Caribes and Taínos) were notable. The latter – settled in the eastern part of the main island – contributed most to Cuba's agricultural heritage, leaving a legacy of plant genetic resources and cultural assets still present today, such as tobacco (*Nicotiana tabacum*), chili peppers and peppers (*Capsicum* spp.) (Pérez Cruz, 2014). According

DOI: 10.4324/9781315183886-13

to Guanche Pérez (2014) the toponym Cuba is of Taíno origins and linked to the concept of garden or orchard. Since the Taínos practiced the slash and burn method of cultivation, the word Cuba assimilates the concept of "cultivated land".

The strategic position of the main Caribbean islands Cuba and Hispaniola and their great potential for food production, was evident for the Conquistadors which they used as operational bases for the subsequent colonization of the continent. New genetic resources were brought by the colonisers to supplement those already present, thus creating an agriculture system capable of sustaining human populations that – in the case of Cuba – have been estimated at about 100,000 inhabitants at the time of the European conquest (Pérez Cruz, 2014). Cuba and the Caribbean made a notable contribution to the early enrichment of European crop plant resources. Different varieties of maize from Cuba contributed to the introduction of the crop to southern European countries, as their genotypic characteristics were more suitable than Mesoamerican varieties (Martínez Viera, 2004). On the other hand, the European conquest brought new genetic resources to the subsistence farm or *conuco*. Asian plants such as sugarcane (*Saccharum officinarum*) and plantain (*Musa* sp.), and the African yam (*Dioscorea* spp.) were integrated with the local Caribbean fauna very early on. The same process took place with domestic animals such as pigs, goats, dogs and sheep, which reached the Antilles from the Canary Islands (Marín-Gutiérrez, 2012).

Following the start of Spanish colonization around 1511, the island remained for more than two and a half centuries with an agricultural economy based on animal husbandry and the cultivation of tobacco to supply external markets, both formally (through the Spanish monopoly) and informally through smuggling. At the same time, agriculture was developed to meet the modest domestic demand (Moreno Fraginals, 1978). The first Cuban rural farmers were essentially aboriginal, and, in this sense, agriculture was an evolution of the Taíno *conuco*. The *conuco* occupied less space than areas dedicated to raising livestock and was generally adjacent to villages as the agricultural produce was intended to be consumed locally.

Livestock systems

Due to the low importance of mining on the island, many settlers engaged in cattle ranching as an extensive production system. Cattle bred faster than land was appropriated, and the establishment of new cattle ranches was often interrupted by the extreme mobility of settlers due to the ongoing conquest taking place on the continent. This is how cattle drives – exploited as a communal asset – emerged spontaneously as a common practice among the first inhabitants of the Spanish colony (Aguilar et al., 2004). In the mid-16th century, cattle raising became the main economic activity on the island and encouraged both by domestic demand and foreign trade. Thus, the production of hides and *tasajo* (dried meat for consumption by shipping fleets) set the basis for a colonial export capitalist economy in Cuba (Moreno Fraginals, 1978).

In the 17th century, livestock still carried significant weight in the island's economy, but the gradual contraction of the shipping fleet system (disappeared in the

18th century), together with the growing importance of other cash crops, caused its importance to decline (Aguilar et al., 2004). Notwithstanding, some breeds of minor livestock species known as *criollos*, genetically linked to zoogenetic resources introduced during the early colonial period, are still bred today. This is the case of the *Pelibuey* sheep, which is widespread on the island, or Creole pigs, with similar origins (Velázquez et al., 1998; Aguilar Martínez et al., 2017). Small livestock (particularly pigs) are historically associated with the *conuco*, given their diet based on by-products from the *conuco*. These animals provide a source of meat and income through commercialization which is an integral component of the subsistence economy of the *conuco*.

The integration of small livestock in agrobiodiverse landscapes where traditional farming practices are applied is found in the western part of the country, such as in the Valle de Viñales where the fruit of the Cuban oak (*Quercus sagrana*) is used as feed. This farming system is culturally linked to similar practices in the Mediterranean *dehesa/montado* where the species *Quercus ilex* and *Quercus suber* (Fra Paleo, 2010) are dominant.

Tobacco in the production system

Tobacco was part of the *conuco* of native peoples such as the Taínos. Canary Island migrant farmers who settled in the colony contributed greatly to its evolution, leading to a process of transculturation. While migration from the Canary Islands – both substantial and sustained – was primarily to rural areas, migration from the Iberian Peninsula was predominantly to urban areas, hence the influence of the Canary islanders on the rural culture of Cuba. These farmers cultivated the tobacco plant in the central part of the island on small plots in natural river meadows. For this reason, Canary islanders in Cuba were named *vegueros isleños* (Marín-Gutiérrez, 2012) highlighting the symbolic difference in names attributed to market agriculture land (*vega*) and farmland dedicated to self-consumption (*conuco*).

Since tobacco is not a perennial crop, this allows crop rotation and temporal variability in agrobiodiversity. Furthermore, fallow land in the *vega* is also important as it can be used to graze livestock. In addition to tobacco, the Canary islanders farmed *conucos* to produce *maloja* (a type of food derived from maize) for livestock, as well as vegetables for family consumption. Thus, a significant volume of small-scale agricultural trade in Cuba was in their hands (Marín-Gutiérrez, 2012).

At present, there is still a predominance of tobacco farming systems associated with other crops and the grazing of small livestock. This increased quality of the landscapes, such as in the Viñales Valley and other locations, and influenced the richness of vernacular architecture with the construction of distinctive buildings used to dry tobacco.

Biodiversity of sugarcane in the conuco

In the first two centuries of the colonial period, sugar played only a secondary role in the production system. Farming was dominated by large tobacco plantations both for export and domestic consumption, followed by animal husbandry and forestry; the latter provided prized woods for shipment to Europe and for the construction of vessels in the shipyards of La Habana. Unexpectedly, the year 1762 marked a turning point for the cultivation of sugarcane. The 11-month long British occupation of La Habana brought with it the trade of greater numbers of slaves and the acceleration of the export economy, which represented an incentive for the creole elite driven by benefits. The onset of the Haitian Revolution in 1791 gave Cuba the opportunity of becoming an alternative sugar producer and exporter due to the instability in the French colony. Nevertheless, land dedicated for self-consumption, or *conuco*, within the sugarcane plantations was farmed by slaves. Indeed, the landowners gave small areas to their slaves for cultivation, which gave rise to the *conuco* of black African slaves who introduced their specific crops, combined with poultry and pig raising. Thus, the owners of the sugar mills would often buy produce from their slaves' *conucos* (Moreno Fraginals, 1978).

Today, varieties of sugarcane previously cultivated intensively for marketing, but no longer used as commercial varieties, can be found in many *conucos*, particularly in mountain regions. Thus, these *conucos* near sugarcane cultivations (old or present) are reservoirs of agrobiodiversity of these varieties which are used by farmers for animal and human consumption, such as for the extraction of its juice (a traditional beverage in Cuba known as *guarapo*) and which is occasionally sold in small quantities.

Coffee and agroforestry systems

The cultivation of coffee – considering the ups and downs of Cuba's agricultural history – is a metaphor for the changes and resilience of both the nation and of its citizens. The introduction of the first coffee plant to Cuba was documented in 1748 (Lapique and García, 2014), and – relying on its impact on international markets – the peak of production did not come until the first half of the nineteenth century when production in the French colony of Santo Domingo collapsed (Moreno Fraginals, 1978).

The agroforestry condition of coffee cultivation favours a high level of agrobiodiversity due to the coexistence with wood species, particularly in mountain regions. Given the rural nature of this system and the duration of the cultivation, traditional practices and knowledge are applied, which increases intraspecific variability (Eyzaguirre and Linares, 2004; García and Castiñeiras, 2006). With the spread of coffee cultivation, both the *conuco* and plantations reached the forest frontier (González Álvarez et al., 2016; Agnoletti et al., 2022). The North American traveller Abiel Abbot (1829) left written testimony in 1829 of the development of this productive system.

> Plantations of coffee, beautifully laid out and neatly cultivated, are almost continuous, and the eye of the traveller is constantly delighted with the finest specimens of agriculture. [...] trees with care will flourish for twenty or thirty years and yield a coffee in higher demand in the market than any other.

The provident planter, however, has taken care to secure to himself more land than he wishes at once to occupy, and every year a crest of hill is reduced from a forest state to plantation, so that while spots in very unfavorable situations become sterile, other new and very fruitful ones are coming on, to preserve complete the full number of fruitful trees.

In addition to land prepared for the cultivation of coffee, corrales or areas for raising animals were also established, attributing greater sustainability to the farms (Ramírez Pérez and Paredes Pupo, 2003). According to Lapique and García (2014), in the western region of Cuba coffee farming is associated with higher agrobiodiversity than agricultural systems based on sugarcane. A quote from Anselmo Suárez y Romero – a Cuban aristocrat cited by the authors – states the following:

I don't like reed beds (…) when you've seen one, you can say that you've seen them all: only immense plantations of green furrowed by guardrails from which the human eye derives no delight from the tops of the caimito or avocado trees, or from the fragrance of oranges and lemons, as in the coffee plantation.

According to Pérez de la Riva (1944), the landscape in agroecosystems based on coffee cultivation was in some cases so valuable that those systems established in the municipalities of Alquízar and Artemisa were considered as "the Garden of Cuba". Coffee agroecosystems including coconut and cocoa cultivation are also found in the Baracoa region, where also small livestock is raised (Sánchez et al., 2015).

The conuco with coffee cultivation as Cuban agricultural heritage

The *Globally Important Agricultural Heritage Systems* (GIAHS) initiative, launched in 2002 at the World Summit on Sustainable Development and later becoming a FAO programme, has the objective of promoting the international recognition, conservation and management of agricultural systems with heritage values (FAO, 2018). The recognition of these sites is based on five criteria: (1) Food and livelihood security; (2) Biodiversity and ecosystem functions; (3) Knowledge systems and adapted technologies; (4) Cultures, value systems and social organizations (agriculture) and (5) Outstanding landscape, and land and water resources sustainable management.

Despite the transformations driven by industrial farming in Cuba's agricultural systems, there still are systems closely linked to local conditions and with relatively well-preserved landscapes. These are maintained by local communities who are bearers of agricultural traditions; which is the case for coffee cultivation which has a relatively low level of technology use as predominantly cultivated in mountain regions. These human settlements are based on subsistence agriculture, with a variety of *conucos* as coffee agroecosystems.

Studies carried out in *conucos* in mountain areas of Cuba, especially in the Sierra del Rosario and Cuchillas del Toa Biosphere Reserves (Watson and Eyzaguirre, 2002; Castiñeiras et al., 2012; Sánchez et al., 2015), have shown that there is a practice of agroforestry, involving a high number of planted and wild trees as well as crops (cultivars and domestic animal breeds). These systems also integrate ornamental and medicinal plants (cultivated and wild), generating supplementary income. Overall, more than 500 plant species with reported uses have been catalogued. These characteristics tie in with two main goals of GIAHS sites: systems that provide food and livelihoods for their inhabitants (1), while preserving biodiversity and ecosystem functions (2).

In *conucos* linked to mountain agroforestry systems – particularly those established by the slash and burn system – farmers tend to cultivate crops for home consumption rarely found in markets. This is the case of the genus *Capsicum* spp., with a high variability associated with traditional and advanced cultivars. Within the diversity of this genus there are semi-domesticated or wild taxa cultivated by farmers, such as the chili bell pepper *Corazón de paloma* found only in one region of Cuba pertaining to the species *Capsicum chinense* (Barrios et al., 2007). Other species in *conucos* are underutilized outside of them, making them critical reservoirs. Such is the case of the *sagú* (*Maranta arundinacea*), an edible tuber, or the lima bean (*Phaseolus lunatus*) which has a high local diversity.

Other meaningful crops in *conucos* associated with coffee cultivation are *malanga* (*Xanthosoma* spp. and *Colocasia* sp.) and cassava (*Manihot esculenta*), or fruit species such as the yellow mamey (*Pouteria sapota*) and pineapple (*Ananas comosus*). This agrobiodiversity is enriched by forest species found in areas adjacent to the *conuco,* which provide shade for the coffee plants, such as the quickstick tree (*Gliricidia sepium*), a supplier of food for pollinators and other animals. Endemic forest species such as *Pinus caribaea* (*pino macho*), as well as wild plant species of varied purposes (timber, medicine, fruit, etc.) in these forest areas are also used by farmers. Endemic mammal species such as rodents of the Capromydae family are frequently hunted for meat: *Capromys pilorides* (*Jutía conga*) and *Mysateles prehensilis (Jutía carabalí).* These values are in line with GIAHS goals of preserving Biodiversity and ecosystem functions (2) and Knowledge systems and adapted technologies (3).

Unique Knowledge systems and adapted technologies (3) in *conucos* linked to agroforestry systems (particularly coffee) are the result of the influence of Afro-Cuban religions in the utilization of biological resources. This traditional knowledge is often observed – linked to the existing diversity – in the production of handicrafts such as furniture, musical instruments, basket weaving and other handicrafts.

Conuco farmers are grouped into Cooperativas de Créditos y Servicios, which organise local and national cooperation as part of the Asociación Nacional de Agricultores Pequeños (ANAP). This strength is salient, since the GIAHS Programme supports communities and local governments in the adoption and development of environmental policies, thus promoting sustainable development, which is an essential principle for the recognition of GIAHS sites [Cultures, value systems and social organizations (Agri-Culture) (4)].

Finally, landscape values unquestionably play an important role in GIAHS sites, [Remarkable landscapes, land and water resources management features (5)]. Mountain *conucos* in forest and agroforestry landscapes have a high diversity of land (González Álvarez et al., 2016), and water uses of environmental and social relevance.

For integrating those values with rural development approaches that maintain the natural and agricultural diversity as a whole, Cuban *conucos* have a significance as aligned with GIAHS, especially those associated with coffee agroforestry systems.

References

Abiel, A. (1829) *Letters written in the interior of Cuba, between the mountains of Arcana to the east, and of Cusco to the west: in the months of February, March, April, and May 1828*, Bowles and Dearborn, Boston, MA

Agnoletti, M., Molina Pelegrín, Y. and González Alvarez, A. (2022) 'The traditional agroforestry systems of Sierra Maestra and Sierra del Rosario, Cuba', *Biodiversity and Conservation*, no 31, pp2259–2296

Aguilar, R., Bu, A., Dresdner, J., Fernández, P., González, A., Polanco, C. and Tasini, R. (2004) *La ganadería en Cuba: desempeño y desafíos*, Editorial Universidad de la República, Uruguay

Aguilar Martínez, C.U, Berruecos-Villalobos, J.M, Espinoza-Gutiérrez, B., Segura-Correa, J.C. Valencia Méndez, J. and Roldán-Roldán, A. (2017) 'Origen, Historia y Situación actual de la oveja Pelibuey en México', *Tropical and Subtropical Agroecosystems*, no 20, pp429–439, https://www.redalyc.org, accessed April 07 2022

Barrios, O., Fuentes, V., Shagarodsky, T., Cristobal, R. Castiñeiras, L., Fundora, Z., Garcia, M., Giraudi, C., Fernandez, L., Leon, N., Hernandez, F., Moreno, V., Arzola, D., Acuña, G., Abreu, S. and de Armas, D. (2007) 'Variabilidad intraespecífica de los recursos genéticos de *Capsicum* spp. conservados en sistemas de agricultura tradicional en Cuba', *Agrotecnia de Cuba*, vol 31, no 2, pp211–219

Carratalá Pérez, A. (2014) 'Los primeros descubridores de Cuba y las Antillas', in F.J. Perez Cruz (ed) *Los Indoamericanos en Cuba. Estudios abiertos al presente*, Editorial Ciencias Sociales, La Habana, Cuba

Castiñeiras, L., Sánchez, Y., García, M., Shagarodsky, T., Fuentes, V., Giraudy, C., Hernández, F. and Hodgkin, T. (2012) 'Oportunidades de Conservar la Biodiversidad Agrícola en las Reservas de la Biosfera de Cuba', *Naturaleza y Desarrollo*, vol 10, no 2, pp18–36

Eyzaguirre, P. and Linares, O. (2004) *Home gardens and agrobiodiversity*, Smithsonian Books, Washington, DC

FAO (2018) *Sistemas Importantes del Patrimonio Agrícola Mundial. La biodiversidad agrícola y los ecosistemas resilientes. Prácticas agrícolas tradicionales e identidad cultural*, https://www.fao.org/3/i9187es/I9187ES.pdf, accessed May 31 2022

Fra Paleo, U. (2010) 'The dehesa/montado landscape (Portugal/Spain)', in K. Belair, I. Ichikawa, B.Y.L. Wong and K.J. Mulongoy (eds) *Sustainable use of biological diversity in socio-ecological production landscapes*, Secretariat of the Convention on Biological Diversity, Technical Series no. 52, Montreal, pp125–127

García, M. and Castiñeiras, L. (2006) *La diversidad agrícola en reservas de la biosfera de Cuba*, Editorial Academia, La Habana, 80pp

González Álvarez, A., Sánchez, Y., Arzola, D., Zamora, J.L. and Hernández, F. (2016) 'Agrobiodiversidad en la Sierra del Rosario, Cuba: el café (*Coffea arabica* L.) y otras claves de su configuración', *Agrotecnia de Cuba*, vol 40, no 2, pp3–8

Guanche Pérez, J. (2014) 'Legado aborigen a la cultura cubana', in F. Pérez Cruz, (ed) *Los Indoamericanos en Cuba. Estudios abiertos al presente*, Editorial Ciencias Sociales, La Habana, pp321–332

Koohafkhan, P. and Altieri, M. (2016) *Forgotten agricultural heritage. Reconnecting food systems and sustainable development*, Routledge, New York, 271pp

Lapique, Z. and García, L. (2014) 'La Periferia Habanera', *Revista Bimestre Cubana*, vol 41, no 2, pp89–102

Marín-Gutiérrez, I. (2012) *Tabaco e inmigración canaria en la región central de Cuba. El caso de la provincia de Santa Clara*, https://www.researchgate.net/publication/280134276, accessed January 05 2022

Martínez Viera, R. (2004) *Estación Experimental Agronómica de Santiago de las Vegas. 100 años de historia*, Ministerio de la Agricultura (MINAG), La Habana

MINAG (2007) 'Informe de país sobre el estado de los recursos fitogenéticos', http://www.fao.org, accessed May 20 2022

Moreno Fraginals, M. (1978) *El Ingenio, Complejo económico-social cubano del azúcar*, Editorial Ciencias Sociales, Tomo I, La Habana, 335pp

Pérez Cruz, F.J. (2014) *Los indoamericanos en Cuba. Estudios abiertos al presente*, Editorial de Ciencias Sociales, La Habana, 512pp

Pérez de la Riva, F. (1944) *El café. Historia de su cultivo y explotación en Cuba,* Jesús Montero, La Habana, 383pp

Ramírez Pérez, J.F. and Paredes Pupo, F.A. (2003) *Los cafetales de la Sierra del Rosario (1790-1850),* Ediciones Unión, La Habana, 103pp

Sánchez, Y., González, A., Kauffman, M., Zamora, J.L. and Pérez, H.M. (2015) 'Family farming and its landscape imbrication. An assessment in Sierra del Rosario and Cuchillas del Toa Biosphere Reserve', *Agrotecnia de Cuba*, vol 5, pp73–85

Velázquez, F., Barba, C., Pérez-Pineda, E. and J.V. Delgado (1998) *El cerdo negro criollo cubano: origen, evolución y situación actual*, https://core.ac.uk, accessed July 4 2022

Watson, J.W. and Eyzaguirre, P.B. (2002) *Home gardens and in situ conservation of plant genetic resources in farming systems*, IPGRI, Rome, 183pp

13 Urban agriculture

A view from Cuba

Noel J. Arozarena Daza, Yanisbell Sánchez Rodríguez, Maribel González-Chávez Díaz and Juan A. Soto Mena

The concept of urban agriculture also includes farming in peri-urban or suburban areas and family farming. This approach, currently in use in Cuba, approximates that reported by Medel (2011) who examines the origin and development of a system that integrates local knowledge, some elements of conventional agriculture as well as agroecological practices and related technologies. Intra and peri-urban areas are used for the production or transformation of agricultural products and livestock, either for sale or direct consumption, taking full advantage of and sustainably re-utilising local resources. Urban agriculture takes place close to the markets where products are generally sold, thus minimising costs for transportation (except in the case of contracted production which requires transportation to the place of sale) and reducing the environmental impact of this activity. It also uses urban resources such as organic waste, thus helping improve the availability of food and the range of biodiverse products consumed by the population. Additionally, it allows greater use of unproductive or underutilised areas, transforming these into productive units which are important sources of employment and earning for families or individuals. To illustrate this, García et al. (2019) identified the use in La Habana, with 19.0% of the national population (ONEI, 2017), of solid waste from more than 300 agricultural markets as raw material to obtain compost and contributing at the same time to the reduction of the ecological footprint. Urban agriculture entails higher complexity and level of integration of land uses. Mougeot (2001) pointed out the need for a greater conceptual maturity of urban farming to adequately align its processes to be recognised as a formal socioeconomic activity. According to this author, for the concept of urban farming to be operational, three factors must be taken into consideration: the difference between rural and urban farming, the complementarity of the latter for the former, and the integration of urban farming in the urban socioeconomic system.

Historical context: the stages of urban agriculture in Cuba

First stage

During the colonial period, from 1492 to 1901, agriculture was incentivised close to inhabited areas due to the scarcity of roads or other communication networks

DOI: 10.4324/9781315183886-14

between territories dedicated to production and centres of commercialisation and consumption (Ortíz, 1991; Acevedo et al., 2014). Despite this, agriculture of this kind coexisted with the growth of large estates elsewhere with sugar cane and live-stock farming, and on a more local scale – in pre-montane and montane regions – to the production of coffee (*Coffea arabica*).

The production of minor crops close to inhabited areas continued up until the 18th and 19th centuries. It is worth noting that Chinese immigration during the 19th century greatly influenced the cultivation of leafy vegetables, and spices both within and in the outskirts of towns. This practice continued for a long time whereas vast and more remote agricultural areas adopted a monoculture system for crops with greater economic value such as sugar cane (*Saccharum officinarum*), coffee and tobacco (*Nicotiana tabacum*), as well as some grains, fruits and *viandas* (a collective term for roots and vegetables such as potatoes, cassava, squash and a wide variety of yams, normally eaten fried or boiled).

The transition from colony to Republic in 1902 brought about changes in the social structure associated with a shift in the political, economic, cultural and legal frameworks. According to Martínez (2018), communication networks, rail transport and technology linked to agriculture received foreign capital investment which significantly stimulated these sectors in addition to the know-how, innovations, technologies, inputs and services arriving from Europe and the United States. Being the base of the national economy (with sugar cane and tobacco as the main commercial crops), the agricultural sector became the scenario for transformation and organisational development which led to greater diversification. An example of this was the creation of the Estación Experimental Agronómica de Santiago de Las Vegas in 1904 in La Habana, now the Instituto de Investigaciones Fundamentales en Agricultura Tropical Alejandro de Humboldt (INIFAT in the Spanish acronym).

The transformation of rural agriculture had the effect of limiting the urban production of food at a low level – but without disappearing entirely – for the first half of the 20th century (Acevedo et al., 2014). García et al. (2014) note that both the land tenure structure (1.0% of owners controlled 50.0% of all arable land) and the export model contributed to create a highly specialised agricultural sector dedicated to the intensive cultivation of a short number of crops which led to food scarcity and a high dependence on external resources (Gonnet, 2018).

Second stage: 1959–1990

In May 1959, the year of the Cuban Revolution, the First Agrarian Reform Law was passed which laid down the foundations for a national agriculture system that was complemented by the Second Agrarian Reform Law of October 1963. 80.0% of arable land in Cuba was nationalised and 20.0% was redistributed to farmers (García et al., 2014). Large farming estates were created and a sustained cooperation between direct and indirect actors involved in agricultural production and in the provision of services and knowledge management started. This led to an intense mechanisation of farms and the utilisation of agrochemicals, giving rise to an agricultural system based on high energy consumption greatly dependent on external inputs, as a response both to the productivist concept of the Green Revolution (Vázquez, 2010; Gonnet, 2018) and to secure sufficient food for the population.

According to Mederos et al. (2014), the consequence of this agriculture model was an increase in food production and quality of life of the Cuban people – particularly of the rural population – which lasted for a number of years. However, the 1980s brought a decrease in the production and efficiency of this agricultural system which led to a growing dependence on food imports, demonstrating the need for organisational and structural reforms in the sector (Arce-Rodríguez, 2012; Gonnet, 2018). According to Herrera (2009), in this period Cuba imported 98.0% of its fuel, 57.0% protein and more than 50.0% of all calories consumed by the population, as well as 97.0% of its livestock feed.

This was the state of agriculture in Cuba at the start of the 1990s when the European Socialism bloc ceased to exist and the USSR – which had until then been the main trading partner for the island nation and guarantor of its fuel supply – collapsed. Deprived of its external resources and its preferential socialist-bloc markets, commercial farming practically collapsed (Mederos et al., 2014), in a scenario in which the gross domestic product (GDP) fell by 35.0%. The sugar cane industry – the country's main agricultural crop – suffered greatly which led to the closure and dismantlement of around 50.0% of the sugar cane processing plants at the start of the 21st century. This entailed an extraordinary effort by the State to accommodate the high number of jobless workers and mitigate the sociocultural impact of that measure.

Third stage: from 1991 to the present

What followed next was a period in which non-mechanised agricultural production was promoted, returning to use of animal traction for farming and transportation, low-input farming practices, the participation of farmer in agricultural practices decision-making, as well as the involvement of the urban population in food production to secure self-sufficiency (Castañeda et al., 2017).

Sustainability as an attribute of agriculture, and agroecology as the conceptual basis for farming were adopted as key working principles. Vázquez (2010), Machín et al. (2010), Funes-Monzote (2016) and Roque et al. (2016) identify the know-how and experience required for an integrated and ecological management of pests, sustainable agriculture, management of biodiversity, dissemination of good agroecological practices between farmers, as well as advances in agroecology as farming science and practice in the country. In this context, urban agriculture became an opportunity to address the food deficit, primarily for vegetables and followed by other crops, caused by the almost total stagnation of the agroindustrial sector. Additionally, it was an opportunity for employment in a scenario of cessation or reduction of economic activity, including part of the sugar agribusiness, which increased the available workforce throughout the country (Acevedo et al., 2014).

Urban agriculture thus became an institutionalised socioeconomic practice and was placed under the coordination of the Ministry of Agriculture (MINAG in the Spanish acronym) through INIFAT, the leading and coordinating institution of the Movimiento de Agricultura Urbana. This Movement was the organisational structure adopted in 1994 (Rodríguez, 2004), currently known as the Programa de Agricultura Urbana, Suburbana y Familiar (AU/ASU/AF in the Spanish acronym) and is established as a Directorate within the MINAG organisation. This programme is basically structured around three directives or main thematic areas: agricultural production, livestock production and support and services activities, within which certain lines of action are prioritised (Herrera, 2009; GNAUSUF, 2020). The Programme is detailed in the Annual Guidelines, which plan all AU/ASU/AF activities each year.

The relevance of urban agriculture for the country was evident in 1998 when the Circular No. 03-98, of the Executive Committee of the Council of Ministers encouraged an accelerated and sustainable production of vegetables in urban areas (González et al., 2008). The AU/ASU/AF was later supported with the promulgation of the Decree-Laws no. 258 (GORC, 2008) and no. 300 (GORC, 2012) which prioritised the incorporation of natural persons in the working of vacant or underutilised land property of the State. This contributed to increasing local food self-sufficiency and improved food security (Pérez et al., 2020), and was complemented with other actions in science, technology and innovation. In this manner, urban agriculture demonstrated its complementary role to rural agriculture, denoted by its inclusion in territorial development plans as a production system (Mougeot, 2001; Medel, 2011; Gobierno de Cuba, 2019).

Today, urban agriculture has an important social role which is also reflected in the actions of the Instituto Nacional de Ordenamiento Territorial y Urbanismo (GORC, 2021), in campaigns aimed at the general public (Dirección Nacional de los CDR, 2021), in its function as a productive system, in its accounting for performance before the Cuban Parliament in several legislatures, and in its inclusion in the country's development strategy until 2030 (SITEAL, 2018). Also, importantly, it is driven by the persisting conditions of supply and resource shortages (Rodríguez, 2004; Bertrán, 2011; Castañeda et al., 2017).

AU/ASU/AF areas and production

Urban agriculture is developed in vacant or underutilised suitable land under the condition that this be used for cultivation, and occasionally for post-harvest processing or marketing activities. It also includes spaces that have changed use or social function, such as patios, attics, domestic gardens, terraces and rooftops.

As for suburban agriculture, this is located in the outskirts of cities involving either smallholdings owned by the farmers or on state-owned land ceded in usufruct for farming (Martínez, 2018). It tends to be more intensive in terms of inputs and oriented towards generating income, and so more labour demanding. Outputs include not only food but also other agricultural supplies in high demand, such as seeds, biological products for pest control, nutritional action or plant growth regulation, fertilisers, animals for breeding or fattening and other products such as charcoal and fodder. The AU/ASU/AF production units usually are less than a hectare. When a larger area is available, it is subdivided into sections by the Ministry of Agriculture to facilitate its utilisation. This is the case of organic farming or organoponics, the most widespread urban cultivation system in the country.

The production from AU/ASU/AF constitutes the basis for small-scale food industry, especially the production of canned or semi-processed foods such as jams, pickles, canned food, wines, desserts and other food products (cakes, pastries, fruit extracts, aromatic beverages, etc.), as well as some non-edible products such as cosmetics, raw materials for artisan production, herbal medicine or active constituents for pharmaceutical use.

Training and the role of the community

Although urban agriculture does not require a higher level of technical qualification, one of the lines of action of the Programme was the instruction of the various stakeholders (GNAUSUF, 2010, 2014, 2020) requiring training in various areas such as the application of phytosanitary measures, the use of agrotechnology, the sustainable management of the soil, as well as the economic and organisational operation of the market. A network of Farm Shop-Offices (*Tiendas-Consultorio del Agricultor*) based in each municipality was developed to facilitate the acquisition of means of protection, tools and supplies such as seeds, seedlings, bioproducts and specialised literature.

Strong links have been built between producers as well as between producers, know-how and research managers under the motto "teach by learning, learn by producing, produce by teaching". Specialists and technical experts of the AU/ASU/AF Programme carry out visits to all municipalities and provinces in the country and provide technical advice to producers and other direct and indirect parties involved in production. Training is also provided through forums and workshops, research-development projects, scientific conferences, courses and seminars, where interaction plays a key role. Postgraduate academic training is also delivered through the master's degree in urban agriculture coordinated by INIFAT since 2005.

González et al. (2008) identified other initiatives such as social projects and community interest groups focused on the exchange and transfer of knowledge on agricultural production and food conservation, for example *Mi Programa Verde* (My Green Programme), a project seeking the reforestation of various areas in La Habana with advice and support from MINAG experts. Also, urban agriculture is practised by prison population in penitentiaries to produce their own food, under the management of the Ministry of the Interior (MININT).

The role of the woman in the AU/ASU/AF

Despite the patriarchal social structure having weakened in Cuban society, there is still much room for improvement (Gonnet, 2018). However, this is not entirely the case for urban agriculture where women represent 47.0% of the employed labour force (Gonnet, 2018). In addition, some collectives have a majority of male workers under female direction and other collectives have a gender bias in favour of women (Arias-Guevara, 2014). For example, Arce-Rodríguez (2012) reports the role of female workers in participatory plant breeding and selection, as bearers of knowledge regarding the preparation, conservation, and consumption of food, and thus are better qualified to select species and varieties based on their organoleptic properties. Many women have been empowered by working in urban agriculture as technical experts, or as owners of local knowledge about certain more specialised activities, such as beekeeping. Still, women have the burden of unpaid household work and family care after their work either in urban or rural areas (Munster and Fleitas, 2014).

With a focus on gender in the Annual Guidelines (GNAUSUF, 2020) of the AU/ASU/AF Programme, various actions have been promoted to drive changes in management and accordingly, synergies with organisations of the Cuban civil society such as the *Federación de Mujeres Cubanas*, the *Comités de Defensa de la Revolución*, the *Central de Trabajadores de Cuba* and the *Asociación Nacional de Agricultores Pequeños* have been laid down.

Knowledge and innovation

The management of innovation or experimentation with the participation of family members has been decisive in the AU/ASU/AF Programme and this helped build identity and a sense of belonging to a social group of urban farmers and contributed to the sustainability of this production system. Vázquez (2010) observed this in his analysis of the transition to agroecology in Cuba. He identifies crop diversification, technologies for sustainable production, organic fertilisation for soil conservation, biological pest control, and the production of biochemical pest control products as areas of experimentation by Cuban farmers, which adds to the innovation practised in the Cuban urban agriculture (Hernández et al., 2010). A feature of innovation and experimentation is a common sequence of stages: generation, validation, adoption, improvement, and the dissemination of results, as manifest in participatory plant breeding (Pino et al., 2005) or the *Movimiento Agroecológico de Campesino a Campesino* (Agroecological Movement Farmer to Farmer) (Machín et al., 2010), among others.

This is aligned with the vision of urban agriculture in Cuba focusing on the development of an agroecological and sustainable farming, the diversification of production, the development of small-scale farming under various forms of land tenure, tailored economic incentives for producers, integration in the urban environment and a citizen's ambition for a healthy diet, all of which may allow the country to reach food sovereignty.

As an institutionalised economic activity (more relevant in times of crisis) urban agriculture accommodated many people who migrated to the city and changed jobs to venture into farming as an alternative due to the stagnation of other economic activities. Their diverse origins and background, however, did not limit their efficacy as food producers. Bertrán (2011) illustrates how about 60% of the urban farmers interviewed in the province of La Habana received no external support or information. This indicates that the lack of work experience in agriculture or academic training in disciplines such as agronomy did not limit innovation by these actors, but rather suggests that they acquired knowledge through practice.

The system of urban, suburban and family farming in Cuba has reached a proven level of organisation validated by its contribution to the food security, by feeding populations with different levels of accessibility to the land, to integrate workers with varied knowledge about agricultural and agroecological practices and turned out to be one of the pillars to attain food sovereignty in the country. Its beneficial impact represents a transcendental milestone in Cuban agriculture for its contribution to.

References

Acevedo, J.A., Gómez, M., López, T. and Díaz, B. (2014) 'Agricultura urbana y periurbana en Cuba', in J. Briz and de M.I. Felipe (eds) *Agricultura urbana, ornamental y alimentaria. Una visión global e internacional*, Ministerio de Agricultura, Alimentación y Medio Ambiente, Madrid, pp323–340

Arce-Rodríguez, M.B. (2012) 'La mujer en la agricultura cubana: recuperación de una experiencia', *Ra Ximhai*, vol 8, no 1, pp127–139

Arias-Guevara, M.A. (2014) 'Género y agroecología en Cuba, entre saberes tradicionales y nuevas tecnologías', *Agroecología*, vol 9, pp23–30

Bertrán, M. (2011) 'La innovación por agricultores en la agricultura urbana en La Habana', Master thesis in Urban Agriculture, Instituto de Investigaciones Fundamentales en Agricultura Tropical Alejandro de Humboldt, La Habana

Castañeda, W., Herrera, A., González, R. and San Marful, E. (2017) 'Población y organoponía como estrategia de desarrollo local', *Novedades en Población*, vol 13, no 25, pp43–55

Dirección Nacional de los CDR (2021) 'Desde el Barrio Cultiva tu Pedacito', Boletín de Orientación e Información a Cuadros y Dirigentes, edición no. 56

Funes-Monzote, F.R. (2016) 'Integración agroecológica y soberanía energética', in F. Funes Aguiar and L.L. Vázquez (eds) *Avances de la agroecología en Cuba*, Estación Experimental de Pastos y Forrajes Indio Hatuey, Matanzas, pp403–420

García, C., Arozarena, N.J., Martínez, F., Hernández, M., Pascual, J.A. and Santana, D. (2019) 'Obtención de compost mediante la biotransformación de residuos sólidos urbanos de mercados agropecuarios', *Cultivos Tropicales*, vol 40, no 2, pp15–33

García, M.E., Mederos, C.M., Fernández, P. and Maestrey, A. (2014) 'Adecuación de políticas, planes y programas para el desarrollo sostenible del sector agropecuario', in Y. Rosaenz (ed.) *Estudio de los factores críticos que inciden en el ciclo de la sostenibilidad alimentaria en Cuba*, Instituto de Investigaciones en Fruticultura Tropical (IIFT), La Habana, pp17–29

GNAUSUF (2010) 'Lineamientos Agricultura Suburbana 2011', Grupo Nacional de Agricultura Urbana y Suburbana, La Habana

GNAUSUF (2014) 'Lineamientos de la Agricultura Urbana, Suburbana y Familiar para el año 2015', Grupo Nacional de Agricultura Urbana, Suburbana y Familiar, La Habana

GNAUSUF (2020) 'Lineamientos de la Agricultura Urbana, Suburbana y Familiar para el año 2020', Grupo Nacional de Agricultura Urbana, Suburbana y Familiar, La Habana

Gobierno de Cuba (2019) 'Cuba: Informe nacional sobre la implementación de la agenda 2030. Objetivos de desarrollo sostenible', Informe voluntario al CEPAL

Gonnet, S.E. (2018) 'Construcciones socioculturales de género en el ámbito rural en Canadá y Cuba. Un estudio comparado', Master thesis, Universidad Central Marta Abreu de Las Villas, Villa Clara

González, M., Castellanos, A. and Price, J.L. (2008) 'Testimonios: Agricultura urbana en Ciudad de La Habana', CIDISAV, La Habana

GORC (2008) 'Decreto-Ley 259 Sobre la entrega de tierras ociosas en usufructo', Gaceta Oficial de la República de Cuba, Año CVI, no 4, pp93–95, 11 julio, La Habana

GORC (2012) 'Decreto-Ley 300 Sobre la entrega de tierras estatales ociosas en usufructo', Gaceta Oficial de la República de Cuba, Año CX, no 45 pp1389–1393, 22 octubre, La Habana

GORC (2021) 'Decreto-Ley 42/2021 De la reorganización del sistema de la planificación física y la creación del Instituto Nacional de Ordenamiento Territorial y Urbanismo', Gaceta Oficial de la República de Cuba, vol CXIX, no 96, pp2888–2890, 24 agosto, La Habana

Hernández, L., Pino, M.A. and Varela, M. (2010) 'Experimentación campesina endógena asociada a la agricultura urbana de las provincias Ciudad de La Habana y La Habana', *Cultivos Tropicales*, vol 31, no 2, pp5–11

Herrera, A. (2009) 'Impacto de la agricultura urbana en Cuba', *Novedades en Población*, vol 5, no 9, http://bibliotecavirtual.clacso.org.ar/Cuba/cedem-uh/20100323071744/Impacto.pdf, accessed March 23 2022

Machín, B., Roque, A.M., Ávila, D.R. and Rosset, P.M. (2010) 'Revolución agroecológica: el movimiento de campesino a campesino de la ANAP en Cuba, cuando el campesino ve, hace fe', ANAP y La Vía Campesina, Habana, Cuba y Jakarta, Indonesia, ANAP/La Vía Campesina/OXFAM

Martínez, E. (2018) 'Uso de la biodiversidad en agroecosistemas suburbanos productores de semilla de frijol común (*Phaseolus vulgaris* L) en San Antonio de los Baños, provincia Artemisa', Master thesis, Instituto de Investigaciones Fundamentales en Agricultura Tropical Alejandro de Humboldt, La Habana.

Medel, J. (2011) 'Agricultura urbana de acción participativa. Un acercamiento metodológico para una intervención social en la recuperación integral de áreas urbanas degradadas', Master thesis in Urbanism, Universidad Nacional Autónoma de México, México, D.F.

Mederos, C.M., García, M.E., Gutiérrez, L., Maestrey, A., Bolumen, S., Guevara, A., Nova, A., Cabrera, R. and Pedraza, J. (2014) 'Efectividad de la gestión del ciclo de la sostenibilidad alimentaria', in Y. Rosaenz (ed.), *Estudio de los factores críticos que inciden en el ciclo de la sostenibilidad alimentaria en Cuba. Factor crítico 2*, Instituto de Investigaciones en Fruticultura Tropical, La Habana, pp31–50

Mougeot, L. (2001) 'Agricultura Urbana: Definición, presencia, potencialidades y riesgos', Cuaderno temático 1, pp1–43, https://docplayer.es/17553184-Agricultura-urbana-definicion-presencia-potencialidades-y-riesgos-luc-j-a-mougeot-1-introduccion.html, accessed May 2022

Munster, B. and Fleitas, R. (2014) 'Equidad vs. inequidad de género en el sector agropecuario en Cuba', Centro de Investigaciones de la Economía Mundial, Universidad de La Habana, La Habana, http://www5.uva.es/jec14/comunica/A_EF/A_EF_9.pdf, accessed May 14 2022

ONEI (2017) 'Anuario Estadístico de Cuba 2016', Oficina Nacional de Estadística e Información', La Habana

Ortíz, F. (1991) *Contrapunteo cubano del tabaco y el azúcar*, Editorial de Ciencias Sociales, La Habana

Pérez, J., Pérez, M.C., González, J., González, M. and Ruiz, A. (2020) *Una nueva forma de gestión para entidades agropecuarias: cadenas de valor agroalimentarias. Estudios de caso*, Editora Agroecológica, La Habana

Pino, M., Dominí, M., Ramírez, A., Hernández, L., Ponce, M., Cálves, E., Terán, Z., Yong, A. and Ríos, H. (2005) 'Aspectos metodológicos a tener en cuenta para la implementación del fitomejoramiento participativo en agricultura urbana', *Cultivos Tropicales*, vol 26, no 3, pp17–21

Rodríguez, A. (2004) 'La agricultura urbana en Cuba. Conceptos y avances', Instituto de Investigaciones Fundamentales en Agricultura Tropical, La Habana, http://repositorio. geotech.cu/xmlui/handle/1234/2060, accessed October 15 2021

Roque, A., Funes-Monzote, F., Vega, L.M., Casas, M., Portuondo, M., Caballero, R. and Díaz, T. (2016) 'Haciendo agroecología', Asociación Cubana de Técnicos Agrícolas y Forestales (ACTAF), La Habana

SITEAL (2018) 'Conceptualización del modelo económico y social cubano de desarrollo socialista. Plan Nacional de Desarrollo Económico hasta 2030: propuesta de visión de la nación, ejes y sectores estratégicos', https://siteal.iiep.unesco.org/sites/default/files/ sit_accion_files/siteal_cuba_0368.pdf, accessed December 13 2022

Vázquez, L.L. (2010) 'Agricultores experimentadores en agroecología y transición de la agricultura en Cuba', in M.A. Altieri (ed) *Vertientes del pensamiento agroecológico: Fundamentos y aplicaciones*, Sociedad Científica Latinoamericana de Agroecología, Medellin, pp229–248

14 Cuba's farming market

Opportunities for development

*Michely Vega León, Luis Sáez Tonacca,
José Puente Nápoles and Maribel
González-Chávez Díaz*

The volume and range of marketed agricultural products are largely conditioned by the production system but also by the market channels in a certain region. As for agricultural production, it depends on a wide range of variables, including -but not limited to- the market demand for food, technological and industrial development, seed quality, climate, soil type, crop management and the skills of farmers. At the same time, public policies and incentives to support the development of agri-food systems are also important. While marketing channels – if not already existing – have to be created keeping in mind gross production, accessibility of agricultural production units to the market, and the location of population settlements in relation to both the market and the production areas. Additionally, factors such as farming practices, culinary and cultural traditions, the nutritional needs of consumers, their food habits and concerns regarding food quality, as well as their purchasing power must also be taken into consideration.

According to Arias (2014), in order to increase productivity, the link between food production units and markets must be strengthened. This may lead to food price stabilization, it encourages investment, and the generated monetary surplus has a multiplier effect when reinvested in the rural economy. In this way, new value chains are created and are integrated into the local food market which contributes to consolidating the farming economy. For example, the establishment of farmers' fairs has facilitated the sale of agroecological and organic products in various localities.

By developing a regulatory framework to protect land ceded through usufruct to natural or legal persons, the Cuban government has helped consolidate cooperativism in agriculture and to improve marketing and food production in urban, suburban, rural areas and *patios* (yards). In addition, it has established mechanisms for contracting production, and regulated capped prices for food. However, connecting farmers in areas with poor accessibility (or with few resources) with markets requires greater understanding of the functioning of the market system in a regional or local context, as well as strategic alliances required to overcome factors limiting access to the market and its proper functioning. The connection between farmers cultivating in the Biosphere Reserves and local markets would allow the commercialization of the rich agricultural biodiversity of these areas and satisfy the growing interest of consumers by enjoying themselves in a natural environment

DOI: 10.4324/9781315183886-15

and consuming value-added agricultural products with differentiated identities and qualities, which generates market value (product, territory and people).

The marketing system of agricultural products in Cuba

In Cuba, agricultural land is managed under various forms of ownership: state-owned (78.90%), cooperative (7.10%) and small owners (14%) (ONEI, 2015a). Different systems of production are used, ranging from state-run companies, the *Unidades Básicas de Producción Cooperativa* (UBPC), the *Cooperativas de Producción Agropecuaria* (CPA), the *Cooperativas de Créditos y Servicios* (CCS), to private farmers. These systems of production account for a total food production (excluding sugar cane) of 7,014.5 Mt (Table 14.1). Production from *patios* and family plots have been excluded from these figures (ONEI, 2015b). By systems of production, the non-state sector contributes between 60.9% and 85.6% to the production of tubers, vegetables, wet paddy rice, beans and fruit trees, among others. However, it only contributes 29.3% to citrus production. Since large farming areas are needed for the cultivation of this crop, production is being managed by state-sector operators.

Several market schemes for the commercialization of crops have been implemented: *Mercados Agropecuarios Estatales* (MAE, state agricultural markets), *Puntos de Venta* (PV, retail stalls), *Mercados Agropecuarios de Oferta y Demanda* (MAOD, supply/demand agricultural markets), *Cooperativas No Agropecuarias* (CNA, non-agricultural cooperatives), and *Mercados Agropecuarios Arrendados* (MAA, leased agricultural markets), with disparate participation in the whole market (Figure 14.1).

Table 14.1 Crop yield according to system of production in 2015 (ONEI, 2015b)

Crop	Total yield (Mt)	Yield according to method of cultivation (%)				
		State-owned	Non-state-owned			
			UBPC	CPA	CCS and private farmers	**Total**
Viandas[a] and other vegetables	5,057.8	12.5	5.2	4.4	77.9	**87.5**
Wet paddy rice	418.0	15.6	20.9	2.6	60.9	**84.4**
Corn	363.0	5.3	5.5	3.6	85.6	**94.7**
Beans	117.6	9.0	7.0	5.6	78.5	**91.0**
Citrus fruits	115.4	62.1	7.4	1.2	29.3	**37.9**
Fuit	942.7	8.0	4.1	2.4	85.6	**92.0**
Total	**7,014.5**					

UBPC, Unidades Básicas de Producción Cooperativa; CPA, Cooperativas de Producción Agropecuaria; CCS, Cooperativas de Créditos y Servicios.

[a] A collective term for roots and vegetables such as potatoes, cassava, squash and a wide variety of yams, normally eaten fried or boiled.

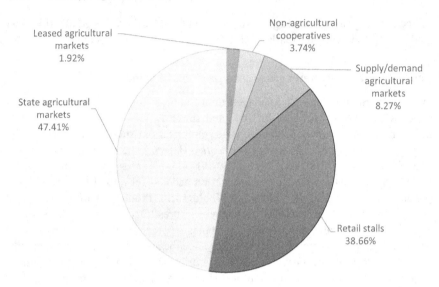

Figure 14.1 Contribution of each type of market scheme to the commercialization of agricultural products in 2015.

Source: ONEI (2015c).

Agricultural products are sold in the MAE with fixed maximum retail prices, while other products are sold according to supply and demand prices and only in the provinces of La Habana, Artemisa and Mayabeque. This market can also be supplied by agricultural companies or by procurement from third parties. Unlike the rest of the provinces, in La Habana, Artemisa and Mayabeque MAEs contract directly with producers and sell through the *Empresa de Mercados de La Habana and the Empresa de Comercio y Gastronomía*. The PV (also known as *kioskos*) are managed by any production unit (state-owned, UBPC, CPA and CCS) with employed personnel, but they can also be managed by private producers (single farmers) that operate within the area. MAOD fix prices based on the balance between supply and demand. MAOD sell products such as potatoes, rice, peas and eggs at capped prices, while all other products are sold at market prices. MAA are rented directly by the units of production (UBPC, CPA and CCS), to sell at market prices, with the exception of those products with capped prices. Other markets for agricultural products can be accessed through direct contracts with social institutions such as day care centres, maternity homes, hospitals, schools, work centres and industries.

The *Unión Nacional de Acopio* (State Procurement Unit) is a public entity responsible for buying products from farmers and distributing them to markets, especially in cities; it contracts more than 3,476 cooperatives each year, with the participation of more than 160,000 agricultural producers. Annually, it dispatches more than 480,000 tons of *viandas*, vegetables, citrus fruits, other fruits and grains (MINAG, 2014). According to data reported by ONEI (2015b), it represents

52.0% of the total production of the country and 47.4% of the whole market. State contracts are critical for the farmer's economy with few resources, living in isolated areas and without means of transport to access the market. The main goals of contracting are to bring production closer to consumers and regulate market prices, thus avoiding increases. In addition to this system, there are parallel or alternative marketing channels (offering more advantageous prices than contracted channels) through which surplus non-contracted products can be sold (directly or through intermediaries), primarily to national tourist services or social entities. When cooperatives and private producers sell directly to the tourism sector, they can generate much higher economic benefits.

Another alternative market are open fairs that take place one weekend a month in all municipalities of the country, to which all cooperatives and intermediaries can participate directly. Contreras et al (2014) observe that open fairs provide a commercial channel for family farmers and artisanal fishermen to sell fruit and vegetable products, fish and seafood, and which is paid for primarily in cash. In addition to promoting local economic circuits, these open fairs are an important institution for the social integration of the population that enables access to healthy food. To promote these fairs, public policies must be established to decentralize marketing and encourage local consumption, while facilitating resources and tools for agricultural production. Actors such as regional and municipal governments are fundamental in the sustainability of these initiatives in rural and peri-urban areas. In the Sierra del Rosario Biosphere Reserve, farmer markets are currently held where seeds are exchanged and agricultural products, ornamental plants, typical dishes, local clothing, or household items as well as other products are sold, with the support of local governments and agricultural cooperatives. Despite these actions, some products remain unsold on the farm in those areas with poor access to markets.

Finally, the nationwide initiative for Urban, Suburban and Family Agriculture (AU/ASU/AF) produces food for local consumption, bringing food closer to population settlements through a network of retail stalls, with a capillary distribution at the neighbourhood level, which has increased the supply and consumption of vegetables locally and reduced the percentage of losses (Vega, 2016). Díaz and Valencia (2014) note that the strengthening of local markets is an undervalued strategy in public policies that should be promoted.

Market opportunities and creation of value chains

According to Andrew (2008), the actions required to connect farmers with markets can take a top-down approach, when identifying a certain market demand and the subsequent search for producers to meet this demand, or a bottom-up approach when selecting the farmers with whom to work and subsequently identifying the markets. Considering the regulation of the national agricultural market and the organizational structure of production and marketing in the country, there are opportunities for the creation and exploitation of value chains and markets.

Some examples are organic or agroecological fruit value chains, local organic or agroecological markets and open fairs to strengthen smaller marketing circuits.

In the Sierra del Rosario and Cuchillas del Toa Biosphere Reserves, the sale of agricultural or processed products predominates, either on an itinerant basis or through formal or informal retail stalls set up along major highways. Some farmers have also developed businesses of processed fruit juices, which they offer to private customers in nearby cities. As there is little contact with local tourism in these areas, neither the offer of such products by local farmers nor the demand are known. On the other hand, there is a shortage of fruits, vegetables, *viandas* and grains in the local markets of the nearest cities. The lack of connection between producers, local markets and tourist facilities, an unawareness of local demand, and low accessibility – due to distance and difficult access roads – have led to high levels of fruit loss and limited cultivation of fresh vegetables. An example is found in the *Cooperativa de Crédito y Servicio Fortalecida Mario Ferro*, located in the Sierra del Rosario mountain range, province of Artemisa, which has 87 members primarily cultivating coffee. These farmers also have groves of the highly marketable *mamey colorado* fruit (*Pouteria sapota* Jacq.) distributed throughout the plantations in addition to other fruit trees that are not marketed due to the lack of contact with markets. However, the province has now developed an industrial capacity to process fresh fruits and supply concentrated juices and jams with guaranteed production volumes to tourist facilities in the area. In this case, the intervention of local actors is necessary to connect production with the demand from the industry sector.

The COBARB project, in alliance with the AU/ASU/AF initiative, managed to connect local farmers from the Yateras municipality with a local market located in the city of Guantánamo. This project generated significant income from the sale of more than 7.23 tons of fruits and vegetables between September 2014 and March 2015 and led to a significant reduction in fruit loss on the farms. Local authorities may foster these markets and adapt them to the local contexts.

Development of a participatory certification system

Production of organic food is an opportunity to expand the economy, generate jobs and create complementary income for small farmers. Therefore, a local Participatory Certification System (PCS) that endorses organic food production would allow farmers to inform consumers that the production area meets certain levels of environmental quality and the farm is managed under agroecological criteria, matching the requirements of tourists visiting the Biosphere Reserves (Gavilanes et al, 2015). Certification of organic farming and processing practices at a local level, based on the evaluation of the farms by the organic producers themselves to verify compliance with national standards followed by validation by external evaluators, is a viable strategy.

The PCS is designed and implemented by the producers and consumers of organic products, which are local stakeholders. PCSs in Cuba operate within the Urban, Suburban and Family Agriculture sector through this programme. Any farmer, however, can take this course of action to certify their production unit. The PCS is an alternative to third-party certification, that acknowledges the practices and know-how of farmers, adds value to products from biosphere reserve areas, and links Family Farming producers to the local food market which is a fundamental component of the four dimensions for food security: availability, access, utilization and stability.

The PCS is based on the voluntary and committed participation of all organic producers (voluntariness and inclusiveness), and on the understanding of the certification mechanisms for organic production. Finally, values and principles of the system must be shared by all stakeholders participating in the process (shared vision). The exchange of knowledge on agroecology among the participants is based on equal opportunities for men and women, generating credibility (participation and gender equity), with the same level of responsibility in certification (horizontality). In this sense, farmer's training should be based on their local experience and know-how by attributing value to knowledge and the use of tools and approaches that promote sustainable local development and the construction of knowledge networks (learning processes). By adopting ethical principles, the PCS also promotes the exchange between institutions, between different sectors and localities, as well as between scientific and local knowledge (integration). Finally, the willingness to provide evidence of farm practices to the certification stakeholders (transparency), provides security and confidence in production.

In the Alejandro de Humboldt National Park, located in the Cuchillas del Toa Biosphere Reserve, six farmers who annually produce around 35 tons of fruit and vegetables in an area of 46.48 ha have joined the system and have been trained as evaluators. The offer for the local organic market is highly varied and includes products such as mango (*Mangifera indica*), coconut (*Cocos nucifera*), avocado (*Persea americana*), guava, sweet orange, sour orange, key lime (*Citrus aurantifolia*), pineapple (*Ananas comosus*), burro banana (*Musa* sp.), sweet potato (*Ipomoea batatas*), cassava (*Manihot esculenta*), tomato (*Solanum lycopersicum*), cucumber, cowpea (*Vigna unguiculata*) and common bean (*Phaseolus lunatus*). In addition, there are non-marketed products such as bija (*Bixa orellana*), Yellow Guinea yam (*Dioscorea* sp.), western cherry laurel (*Prunus occidentalis*), Malay apple (*Syzygium malaccense*), mamoncillo (*Melicoccus bijugatus*), soursop (*Annona reticulata*), cherimoya (*Annona cherimola*) and anón (*Annona squamosa*), with high potential to be marketed.

References

Andrew, W.S. (2008) 'Cómo vincular a los productores con los mercados. Experiencias hasta la fecha', Documento ocasional del Servicio de Gestión, Comercialización y Finanzas Agrícolas 13, FAO, Rome

Arias, P. (2014) 'Marco Teórico: Acceso de la agricultura familiar a las cadenas de valor', in S. Salcedo and L. Guzmán (eds), *Agricultura Familiar en América Latina y el Caribe. Recomendaciones de políticas*, FAO, Rome, pp312–319

Contreras, R., Krivonos, E. and Saez, L. (2014) 'Mercados locales y ferias libres: El caso de Chile', in S. Salcedo and L. Guzmán (eds), *Agricultura Familiar en América Latina y el Caribe. Recomendaciones de políticas*, FAO, Rome, pp369–388

Díaz, T. and Valencia P. (2014) 'Lineamientos para el fortalecimiento de la producción pecuaria familiar en América Latina y el Caribe', in S. Salcedo and L. Guzmán (eds), *Agricultura Familiar en América Latina y el Caribe. Recomendaciones de políticas*, FAO, Rome, pp163–173

Gavilanes, P., Vega, M., Pérez, J. and Gómez J. (2015), *Manual del Sistema Participativo de Garantía (SPG) en Cuba*, Editorial Agroecológica, La Habana

MINAG (2014) 'La comercialización de los productos agrícolas. Estudio de caso: Empresa Acopio Matanzas', *Boletín el productor*, vol 5, no 4, pp8–12

ONEI (2015a) 'Panorama uso de la tierra Cuba', Oficina Nacional de Estadísticas e Información, https://1library.co/document/q0gkdxgz-panorama-uso-de-la-tierra-cuba.html, accessed August 15 2022

ONEI (2015b) 'Sector agropecuario indicadores seleccionados. Enero-diciembre de 2015', Oficina Nacional de Estadísticas e Información, https://cubanosporelmundo.com/2016/05/29/autoridades-cubana-baja-produccion-agropecuaria/, accessed August 12 2022

ONEI (2015c) 'Venta de productos agropecuarios. Indicadores seleccionados', Oficina Nacional de Estadísticas e Información, https://1library.co/document/qmjl45p4-ventas-productos-agropecuarios-indicadores-seleccionados-enero-septiembre.html, accessed November 25 2022

Vega, M. (2016) 'Cuba', in FAO (ed) *Pérdidas y desperdicios de alimentos en América Latina y el Caribe, Boletín 3*, Rome, FAO, p17

15 Valorisation of products and services from Biosphere Reserves of Cuba

Maribel González-Chávez Díaz, Michely Vega León, Luis Sáez Tonacca and Yanisbell Sánchez Rodríguez

The agricultural biodiversity (ABD) in Cuba's markets is limited, both in species and in infraspecific variability, which has caused a decline in the customary use of much of the country's biodiversity and consequently, the loss of culinary traditions. This has had a negative impact in the conservation of native genetic resources and slowed the improvement of the quality of life of the population.

Although legal mechanisms exist that aim to create links between farmers and markets, these are still inadequate. Currently, the population does not have much possibility to choose the foods to consume (including those in season), due to the limited offer in the markets. The current challenge is to promote the valorisation of services and ABD crops grown and preserved by farming families from generation to generation for their own subsistence. Identifying innovative mechanisms to attribute value to this diversity can be the basis for effective sustainable development for rural communities and the population in general, limiting the reduction of native genetic resources and salvaging traditional Cuban culinary culture through the recovery of local knowledge. This would be an important contribution to improve the standard of living of the population and achieve food sovereignty in the country.

Biosphere Reserves are the ideal scenario for ABD conservation which farmers consider an integral part of their culture, traditional knowledge and practices. Previous studies conducted in Cuba (Castiñeiras et al., 2006, 2012) revealed that the Sierra del Rosario and Cuchillas del Toa Biosphere Reserves are refuges for unique components of agricultural biodiversity. Examples include species such as *Benincasa hispida* (*calabaza china*), *Luffa acutangula* (*chiquá*), *Cyperus esculentus* (*chufa*), *Sechium edule* (*chayote*), as well as primitive cultivars of *Capsicum* spp. (Barrios et al., 2007, 2017), *Zea mays* and *Musa* sp. (Shagarodsky and Castiñeiras, 2013), found in those areas.

For this reason, Biosphere Reserves as areas where sustainable use of resources, ecosystems and landscapes is promoted, can host certain the production of goods and services, as well as their commercialisation. This would contribute to diversified supply to consumers and the increase in family income from a sustainable production.

DOI: 10.4324/9781315183886-16

Valorisation of Biosphere Reserve products

The ABD inventory in the Sierra del Rosario and Cuchillas del Toa Biosphere Reserves includes 732 species cultivated by farmers (UNEP/GEF, 2016), of which 73 products (corresponding to 70 species) are currently marketed. Classified by type and for human consumption and use, these species are as follows: 32 species of fruit, 8 species of grains, 12 species of root vegetables and 21 species of leaf vegetables (Table 15.1). Three of these species have more than one form of consumption, e.g., *Phaseolus lunatus* and *P. vulgaris*, which are consumed both as a pulse and vegetable, while *Musa* sp. is consumed as a vegetable and fruit.

Some of these species or their intraspecific variability have limited population levels and are not used except in the location where they are cultivated, and primarily for local consumption; these do not reach the market – even local ones in more urbanised areas – and so are considered as underutilised species and cultivars. This group of species also includes *Phaseolus lunatus*, *Pouteria campechiana*, *Arracacia xanthorrhiza* and *Artocarpus altilis*, among others. An example of underutilised cultivars with very low population levels is the *Manzano* and *Ciento en Boca* bananas (consumed as fruits) with singular quality and flavour but risking genetic erosion.

In the home gardens and plots in these Reserves, some species – underutilised in the rest of the country – are also cultivated for their medicinal use, such as manzanilla (*Phania matricarioides*), marigold (*Calendula officinalis*), passionflower (*Passiflora edulis*), carpenter bush (*Justicia pectoralis*), narrow-leaf plantain (*Plantago lanceolata*), ginger (*Zingiber officinale*) and Java tea (*Orthosiphon aristatus*), among others. These species are difficult to cultivate and propagate in the plain as their optimal growing conditions are in the lower temperatures found at 150 m above sea level. The increased market price of medicinal plants in recent years has incentivised farmers to cultivate them, leading to increased production in certain areas to sell these products to the state and to an increase of the coverage of underutilised species.

Marketing develops in a context in which state commerce constitutes the backbone of the country's economy to secure the food security of the population whereas private commerce (only recently legal in Cuba, between 2010 and 2011) is still in its early phases. For this reason, the valorisation of agricultural products from family farms and their utilisation should be addressed by both the state and private sectors.

Multiple stakeholders are linked to the marketable assets in every territory that play various roles in promoting those underexploited farm products and services. These include not only farmers, but also artisans, local, municipal and provincial governments, non-governmental organisations (NGOs), universities and government departments (including the Ministerio de Ciencia, Tecnología y Medio Ambiente, Ministerio de Agricultura, Ministerio de Educación y Turismo, la Empresa Comercio y Servicio (Universal), Empresa de Gastronomía Especializada, Empresa de Conservas y Vegetales, el Programa Nacional de Agricultura Urbana, Suburbana y Familiar (AU/ASU/AF), as well as small private enterprises dedicated to the

Table 15.1 Main ABD products in the Sierra del Rosario and Cuchillas del Toa Biosphere Reserves with marketing possibilities

Common name	Scientific name	Common name	Scientific name
Fruits		**Grains**	
Aguacate	*Persea americana*	Ajonjolí	*Sesamum indicum*
Albaricoque	*Prunus armeniaca*	Frijol caballero	*Phaseolus lunatus*
Anón	*Annona squamosa*	Frijol caupí	*Vigna unguiculata*
Caimito	*Chrysophylum cainito*	Frijol mungo	*Vigna radiata*
Calabaza china	*Benincasa hispida*	Frijol común	*Phaseolus vulgaris*
Canistel	*Pouteria campechiana*	Frijol gandul	*Cajanus cajan*
Chirimoya	*Annona cherimola*	Maíz	*Zea mays*
Ciruela	*Prunus occidentalis*	Maní	*Arachis hypogaea*
Coco (dry and green)	*Cocos nucifera*		
Fresa	*Fragaria vesca*	**Vegetables**	
Fruta bomba	*Carica papaya*	Acelga	*Beta vulgaris*
Guanábana	*Annona reticulata*	Achicoria	*Cichorium intybus*
Lima	*Citrus aurantifolia*	Ají cachucha	*Capsicum chinense*
Lima dulce	*Citrus medica*	Ajo de montaña	*Allium sp.*
Limón francés	*Citrus jambhiri*	Ajo puerro	*Allium porrum*
Mamey colorado	*Pouteria sapota*	Berenjena	*Solanum melongena*
Mamey de Santo Domingo	*Mammea americana*	Berro	*Nasturtium officinale*
		Brócoli y coliflor	*Brassica oleracea*
Mamoncillo	*Melicoccus bijugatus*	Cebolla multiplicadora	*Allium ascalonicum*
Mandarina	*Citrus reticulata*		
Mango	*Mangifera indica*	Cebollino	*Allium schoenoprasum*
Mapén	*Artocarpus altilis*	Espinaca	*Spinacia oleracea*
Marañón	*Anacardium occidentale*	Habas lima	*Phaseolus lunatus*
Naranja agria	*Citrus aurantium*	Habichuela corta	*Phaseolus vulgaris*
Naranja dulce	*Citrus sinensis*	Lechuga	*Lactuca sativa*
Níspero	*Eriobotrya japonica*	Pepino	*Cucumis sativus*
Piña	*Ananas comosus*	Pimiento	*Capsicum annuum*
Plátano fruta	*Musa sp.*	Quimbombó	*Abelmoschus esculentus*
Tamarindo	*Tamarindus indica*		
Toronja	*Citrus paradisi*	Rábano	*Raphanus sativus*
Zapote	*Manilkara zapota*	Remolacha	*Beta vulgaris*
		Zanahoria	*Daucus carota*
Viandas			
Afió	*Arracacia xanthorrhiza*	Taro/Malanga[a]	*Colocasia esculenta*
Árbol del pan	*Artocarpus altilis*	Malanga de chopo	*Xanthosoma sagittifolium*
Boniato	*Ipomoea batatas*		
Chayote	*Sechium edule*	Ñame	*Dioscorea sp.*
Cúrcuma	*Curcuma longa*	Plátano vianda	*Musa sp.*
Mariposa	*Hedychium coronarium*	Sagú	*Maranta arundinacea*
		Yuca	*Manihot esculenta*

[a] Both names are used for the same species.

production of fresh or processed food. The *Instituto Cubano de Radio y Televisión* (Ministerio de Cultura) might also play an important role in promoting the utilisation of native agricultural biodiversity by disseminating the values of different species and varieties, as well as traditional recipes which today are practically unknown to the country's population.

Gastronomic-cultural festivals in the Sierra del Rosario and Cuchillas del Toa Biosphere Reserves are periodically held with the support of local governments (UNEP/GEF, 2016) to promote local gastronomy using traditional species and cultivars. This has given added value to local farm products not only for the consumers but also at the state level (government departments and market managers) by supporting the marketing of these products.

The presentation of traditional recipes at exhibitions and cooking demonstrations held at fairs in the Reserves, documented with brochures acknowledging the people (name and residence) that traditionally prepare these dishes (Piña et al., 2016), helps increasing the awareness of traditional utilisation of ABD products as well as the self-esteem of families, especially women.

Access to the mountainous areas of these Reserves is problematic; the roads leading to the nearest markets are in poor condition, especially in the Cuchillas del Toa Biosphere Reserve. Thus, when assessing potential marketing benefits it should be taken into account that the farmer's profit should pay for the low-accessibility costs since, otherwise the activity would not be profitable for a rural smallholder. Even with this apparent disadvantage farmers manage to access urban areas, e.g. two farmers in the Cuchillas del Toa Biosphere Reserve (part of the *Cooperativa de Crédito y Servicio (CCS) Lino Álvarez de las Mercedes*) have connected with the markets *Mercado Agropecuario Especializado (MAE)* y *Mercado Climatizado de Guantánamo* (ONEI, 2015). This was the first time that agricultural products cultivated and processed in Las Municiones community (Yateras) began to sell directly in the city of Guantánamo. This activity has contributed significantly to their family income (Table 15.2).

In the same Reserve, a farmer managed to improve his income by 15% (2012–2014), through links with the Specialised Market (MEC) in the city of Guantánamo by increasing the range of products with new species and cultivars from his farm. These products, sweet oranges, limes, sapodilla, bananas, yams and other products, were not subject to State contracts but sought after by the urban population. Additionally, other farmers currently market more than thirty agricultural products and natural fruit juices from their farms through various retail points. Another couple of farmers managed to sell fruit juices (processed in their own mini-industrial plant) to two restaurants in the city of Guantánamo generating significant economic benefit for their families.

In 2016, Hurricane Matthew struck the Baracoa area, causing significant damage in general and the loss of crops resulting in considerably lower yields that year, but the farmers were able to recover. Due to the high quality of the soil, together with proper management practices on their farms (Socorro et al., 2014), they were able to plant short-cycle vegetables and a variety of fruit trees on farms where only coconut cultivated fields had survived.

Table 15.2 Products valorised and sold by the farmers of the Lino Álvarez de las Mercedes CCS, Yateras district (Cuchillas del Toa Biosphere Reserve) in the Mercado Climatizado de Guantánamo (2015)

Product	Quantity (qq)	Sale price (CUP/qq)	Total earning by farmer (CUP)
Sweet orange	89.00	70.00	6,230.00
Bitter orange	12.00	30.00	360.00
Lime	6.23	30.00	186.90
Dessert banana	34.75	45.00	1,563.75
Naseberry	7.50	90.00	675.00
Avocado	16.00	100.00	1,600.00
Yam	5.22	120.00	626.40
Total	**170.70**	–	**11,242.40**

Note: CUP: Cuban peso. Cuban quintal (qq) = 100 libras = 46 kg.

The increase in farmers' income is an example of the resilience of these systems (Kaufmann, 2014; PAR, 2016) and of the implementation of knowledge acquired during the training workshops held with the communities in the Cuchillas del Toa Biosphere Reserve (UNEP/GEF, 2016) on adaptive agroecological practices. The data reported by fourteen farmers from the Baracoa area showed a gradual increase in their family income in the years when marketing activities were being promoted by the COBARB project (UNEP/GEF, 2016). The farmers underline that after that they learned to valorise the products of their farms. Two roadside stalls were opened and used by several farmers to introduce fresh or processed food into the market, including some varieties underexploited by themselves, such as sour orange juice, water, coconut oil and some fruits such as mango and guava (fresh and processed).

In the case of the Sierra del Rosario Biosphere Reserve, some farmers have made direct links to different markets (MECs) in La Habana by marketing their farm-produced tomatoes as a new choice that has generated additional economic benefits. Table 15.3 shows the results of tomato sales in 2017. These farmers had managed to extend the shelf life of their product by making conserves but despite this, a good part of the harvest was lost so the link with the MEC helped reduce this loss considerably.

As observed Henderson (2019), traditional ABD products satisfy the want of genuine taste as well as providing a healthy diet. In the case of the Sierra del Rosario and Cuchillas del Toa Biosphere Reserves, the traditional agricultural biodiversity has high potential to provide products registered as part of Cuba's Participatory Certification System (PCS), strengthening the capacity to promote good agricultural practices to ensure quality control of agricultural products (Vega and Gavilanes, 2016), which must be further promoted.

Valorisation of services in the Biosphere Reserves

Natural and agricultural landscapes, traditional cultural practices and artisanal processing of products are resources suitable to be valued through the tourism industry and

Table 15.3 Tomato sales in La Habana markets by farmers
from Sierra del Rosario (2017)

Market	Quantity (kg)	Total price (CUP)
El Trigal wholesaler	10,363	171,000
Egido	1,818	27,000
Other state markets	4,181	50,600
Total	**16,363**	**248,600**

may complement and diversify the economies of many rural families (Blanco, 2010). Farming can meet the demand of certain tourism segments – particularly, environmental tourism – interested in learning about rural culture and the countryside as well as fostering forms of coexistence by maintaining environmentally sustainable agricultural practices (Rodríguez, 2019). Agroecotourism is a service with a high potential for valorisation in the Biosphere Reserves, and in addition, tourists have the possibility of greater exchange with local residents, especially when according to Fernández (2016), the greatest attraction for tourists is the level of creolism in these families.

Agroecotourism is already being developed in the Sierra del Rosario Reserve, varying with the agro-biological landscape, providing a healthy alternative by offering agrobiodiversity services and products in the form of fruits or natural juices, as well as foods made with traditional recipes, which cannot be found in other tourist infrastructures. Still, the range of services and number of farmers involved should be increased based on the characteristics of each farm. There are tourist circuits currently operative, e.g. Sierra del Rosario Ecological Station – El Rocío Farm – Las Terrazas Tourist Complex and Sierra del Rosario Ecological Station – Eloísa Bocourt Farm – Orquideario de Soroa – Patio Cultural La Montaña y Yo that are a source of income to the families involved (González-Chávez et al., 2021).

The cultural and natural values of the Cuchillas del Toa Biosphere Reserve are considered the most symbolic attraction due to the history and traditions of the local population (González-Chávez et al., 2016). There is a wide range of agroecotourism services offered in the area of Baracoa of the Reserve where an agroecological trail (Recreo-Santa María) follows locations with biocultural assets in four farms managed by rural families. These farms are located in pre-mountain and mountainous areas. Traditional agricultural practices allowing the conservation of native ABD crops, activities related to livestock husbandry as well rural architecture are all key elements.

References

Barrios, O., Fuentes, V., Garcia, M., Castiñeiras, L., Fundora, Z., Sanchez, Y., Gonzalez-Chavez, M., Giraudi, C. and Guzman, F. (2017) 'Rescate y conservación *in situ* de la diversidad silvestre de *Capsicum* spp. (ajíes) en Reservas de la Biosfera de Cuba', *Agrotecnia de Cuba*, vol 41, no 2 pp72–82

Barrios, O., Fuentes, V., Shagarodsky, T., Cristobal, R. Castiñeiras, L., Fundora, Z., Garcia, M., Giraudi, C., Fernández, L., Leon, N., Hernandez, F., Moreno, V., Arzola, D., Acuña, G., Abreu, S. and de Armas, D. (2007) 'Variabilidad intraespecífica de los recursos genéticos de *Capsicum* spp. conservados en sistemas de agricultura tradicional en Cuba', *Agrotecnia de Cuba*, vol 31, no 2, pp 327–335

Blanco, M. (2010) 'Desarrollo de los agronegocios y la agroindustria rural en América Latina y el Caribe. Conceptos, instrumentos y casos de cooperación técnica', Instituto Interamericano de Cooperación para la Agricultura (IICA), Turrialba, 270pp. ISBN 978-92-9248-193-3

Castiñeiras, L., Barrios, O., Fernández, L., León, N., Cristóbal, R., Shagarodsky, T., Fuentes, V., Fundora, Z., Moreno, V., de Armas, D., Acuña, G., García, M., Hernández, F., Arzola, D. and Giraudy, C. (2006) *Catálogo de Cultivares Tradicionales y Nombres Locales en Fincas de las Regiones Occidental y Oriental de Cuba: Frijol Caballero, Frijol Común, Ajíes – Pimientos y Maiz*, Agrinfor, La Habana, 63pp

Castiñeiras, L., Sánchez, Y., García, M., Shagarodsky, T., Fuentes, V., Giraudy, C., Hernández, F. and Hodgkin, T. (2012) 'Oportunidades de Conservar la Biodiversidad Agrícola en las Reservas de la Biosfera de Cuba', *Naturaleza y Desarrollo*, vol 10, no 2, pp18–36

Fernández, J.A. (2016) *Turismo desde el surco: Agroecoturismo cubano*, http://www.cubainformation.tv, accessed March 23 2021

González-Chávez, M., Sánchez, Y., Ortiz, L., Fernández Granda, L., González, A., Begue, G., Guarat Planche, R. F., Soto Mena, J.A. and Rodríguez Díaz, Y. (2016) 'Mapeo de activos y actores para la implementación de una estrategia de desarrollo territorial rural con identidad cultural en Baracoa', *Agrotecnia de Cuba*, vol 40, no 2, pp74–86

González-Chávez, M., Sánchez, Y., Vega M., Ortiz, L., Tejeda, G., González, A., Rodríguez, L., Fernández, L., Moreno, V., Arzola, D., Zamora, J.L., Guarat, R. F., Rodríguez, G., Soto, J.A. and Rodríguez, Y. (2021) 'Agroturismo como alternativa para la valorización de la agrobiodiversidad en la Reserva de Biosfera Sierra del Rosario, Cuba', *Agrotecnia de Cuba*, vol 45, no 2, pp67–74

Henderson, E. (2019) 'Participatory *guarantee systems: wha*t they are, why should we consider them part of our Plan B. The Natural Farmer', Northeast Organic Farming Association, https://www.thenaturalfarmer.org/article/participatory-guarantee-systems/, accessed December 10 2021

Kaufmann, M.C. (2014). 'Maintaining Agrobiodiversity in Man and the Biosphere (MaB) Reserves in Guantánamo, Cuba', Master thesis, University of Applied Sciences, Bern

ONEI (2015) 'Venta de productos agropecuarios. Indicadores seleccionados', Oficina Nacional de Estadísticas e Información, https://1library.co/document/qmjl45p4-ventas-productos-agropecuarios-indicadores-seleccionados-enero-septiembre.html, accessed July 25 2002

PAR (2016) *Landscapes for agrobiodiversity: agrobiodiversity perspectives in land-use decisions*, Platform for Agrobiodiversity Research, https://static1.squarespace.com/static/5cd00f2c94d71a802807347f/t/5ec0ea54cf008f288eeeebb0/1589701262260/landscapes-for-agrobiodiversity-PAR.pdf, accessed December 13 2021

Piña Cordero, A., Rodríguez, Y., Fundora, Z., Fernández, L., Sánchez, Y., Arzola, D., González, Y.C. and Tellaría, T. (2016) *Recetas tradicionales de Sierra del Rosario*, https://isbn.cloud/9789597223191/recetas-tradicionales-de-sierra-del-rosario/, accessed December 11 2022

Rodríguez, G. (2019) 'El Agroturismo, una visión desde el desarrollo sostenible', *Centro Agrícola*, vol 46, no 1, pp62–65

Shagarodsky, T. and Castiñeiras, L. (2013) 'Especies de plantas subutilizadas en Cuba', *Agrotecnia de Cuba*, vol 37, no 1, pp18–25

Socorro, A., Kaufmann, M., González, A., Hernández, Y., Sánchez, Y., Ortíz, L., Cristóbal, R. and Zamora. J.L. (2014) 'La capacidad agrícola de fincas de la Reserva de la Biosfera Cuchillas del Toa, Municipio Baracoa', *Agrotecnia de Cuba*, vol 38, no 1, pp34–44

UNEP/GEF (2016) 'COBARB_Cuba_Final Report-4158 (PIR). Agrobiodiversity Conservation and Man and the Biosphere Reserves in Cuba: Bridging Managed and Natural Landscapes', GEF ID: GFL-2328-2715-4C72, https://www.coursehero.com/file/116688978/4158-2016-PIR-UNEP-Cubadoc/, accessed January 21 2022

Vega, M. and Gavilanes, P. (2016) 'Los Sistemas Participativos de Garantía (SPG), una alternativa para la valorización de los productos de las reservas de la biosfera', *Agrotecnia de Cuba*, vol 40, no 2, pp87–93

16 Voices of the farmers

Representations of sustainable farming and agrobiodiversity

*Celia Cabrera Ibáñez, Alejandro González
Álvarez, Lázaro Lorenzo Ravelo and
Zoraida Mendive*

Over the course of time, farmers in the Biosphere Reserves of Sierra del Rosario and Cuchillas del Toa have built a living legacy of traditional agricultural knowledge linked to agrobiodiversity. Therefore, it is necessary to examine their vision as conservationists, to understand the significance of agricultural biodiversity in both sites and its evolution to preserve it effectively.

By attributing value to the social and cultural dimensions of agrobiodiversity, field research with the aim to learn about farmer's perception and views on agrobiodiversity in the two areas was conducted adopting an ethnographic approach. Meneses Cabrera and Cardozo Cardona (2014) consider that ethnographic methods allow for comprehending the behaviour of social groups and the meanings of actions and interactions taking place in their environment and, for this purpose, they suggest the interview as a suitable method.

The compilation of narratives of various farmers in the reserves – especially those with a higher level of interaction with the communities (as formal or informal leaders) as a whole – as recommended by Narayan and Patsch (2002), provides valuable information, that when complemented with non-participant observation allows the identification of traditional knowledge. This chapter is based on the analysis of 170 semi-structured interviews administered to farmers in Sierra del Rosario (80) and Cuchillas del Toa (90), conducted within the framework of the COBARB project. The main ideas conveyed by the farmers' narrative are organized according to the key topics proposed.

Agrobiodiversity according to farmers

"It is very important to conserve a group of endangered species, both plants and animals", says Juan, a farmer of Los Tumbos in the RBSR. Juan Carlos, from Soroa, another locality in this Reserve refers to agrobiodiversity as *"the most important element for achieving sustainable and environmentally friendly agriculture"*. Differently, Eloísa – a rural farmer and teacher for many years – defines the agrobiodiversity of her farm in Soroa as *"all the variety of crops we have and the living beings that are related to these crops..."* and expresses with concern: *"if we do not conserve agricultural diversity, then we will be the main losers"*.

DOI: 10.4324/9781315183886-17

It is a fact that farmers have embraced the importance of the conservation of diversity from a more conceptual perspective, although from a more implicit perspective this issue is not new. Both reserves have traditionally considered diversification as an economic strategy. *"The mouse that has only one den is lost"*, notes Margot from the RBCT. The Reserves' agroecosystems became more diverse throughout the agricultural history of the rural communities. *"This farm belonged to my grandparents and, since then, we have grown coffee"*, Segundo tells us on his farm La Comadre, in the RBSR. Certainly, in the mountain areas of the reserves, coffee has been pivotal. Indeed, the ecological conditions for the cultivation of coffee are found in these regions whereas sugarcane expanded as the main economic crop (Moreno Fraginals, 2014) in the plains. Jorge – not far from Segundo's farm – conveys the contribution of coffee to the agrobiodiversity of his farm: *"Our fruit trees produce mamey colorado, caimito, cupcake fruit, mango, guava, banana, all kinds of citrus [...] and then timber: majagua, júcaro, cedar, and royal palm, which I use for shade for my coffee!"*.

Agricultural diversification at the farm level is not only a means for survival, but is increasingly meaningful as an option, both for family and local development, in the perspective of new marketing opportunities. It is true that family and traditional agriculture in Cuba is mostly practiced in scenarios little conducive to high productivity. However, farming in marginal areas benefits from an agricultural diversity with a unique productive and cultural character (González Álvarez et al., 2016).

Relationship between agrobiodiversity and conservation

Family farmers are aware of the need to use and maintain ecosystem services to increase agricultural productivity. This is a fundamental premise of agroecology that shapes the culture of rural farmers in different landscapes. Nestor's farm, in Cayajabos (RBSR), is located in a flat area with moderate elevations. There is a strong animal husbandry tradition in the area. This farmer diversifies by integrating livestock and agriculture in a small vegetable garden near a stream that he uses for irrigation: *"Birds, such as cattails, are excellent and great allies in eliminating skin parasites in cattle. And in the stream – which I keep clean – I even have shrimp"*.

Several professionals working in the Biosphere Reserves are also farmers, which makes them fitting go-between to link agrobiodiversity practice to the institutional perspective. They are also key stakeholders to gather local knowledge used in management and decision-making, an aspect emphasized by Beckford and Barker (2007). Protected areas are rich in agroecosystems that make agricultural production compatible with the conservation of biodiversity and ecosystem services. This type of farmer more deeply recognizes diversity as a source of opportunities. This is the opinion of Lucho, an entrepreneur and farmer in Vega Grande (RBCT), and at the same time an employee of a governmental department to which the Reserve belongs: *"When working on the farm you have to have a keen eye. In nature everything works"*. Isidro, an experienced farmer from La Comadre (RBSR), indicates that: *"The forest helps a lot to mitigate drought, it doesn't let the sun hit the land hard and there is less soil loss when the rains come, which is a great help"*. According to Lele, from Soroa (RBSR), *"the best way to conserve natural resources is to love the land, love to work on it, and most of all to avoid using chemicals, so that all the products are natural"*. Local knowledge also includes biological pest control, as expressed by Margot (RBCT): *"The lion ant is very good because it keeps the leafcutter that kills the crops under control, and also the weevils that spoil the sweet potato. When I see one, I'm really happy!"*.

Agricultural management practices

In the Sierra del Rosario and Cuchillas del Toa Biosphere Reserves, farmers have adapted their management practices to the type of farm and location, so that empirically they find appropriate ways to manage their agroecosystems, as Segundo (RBSR) indicates:

> ... it is the land that tells you... I try out poor land, planting some seedlings around, a seedling here and another over there... and in time I know where and what I am going to plant the following year. Look, I plant cassava in higher areas to protect it from rain and floods.

Management practices have been handed down from one generation to the next, as confirmed by Lele (RBSR): "I learned almost everything I know about working on the farm from my parents. I picked up my experience by watching them till the land". Crop association is one of the practices adopted by farmers to maintain a balance in soil nutrients, as well as for natural prevention of pests: "I plant maize together with beans and squash...", says Arcadio (RBSR), who also reveals that

> okra with beans makes a barrier so that pests don't get through. I plant maize and beans together because apart from having to prepare the soil just once, weeds and pests are kept in check and they both grow well.

Manuel, from Carambola, also in the RBSR, comments: "It is essential to have a correct crop rotation and to prioritize polyculture on the farm as well as carrying out adequate soil management with biological alternatives that allow it to be fertilized and prevent it from being eroded".

Another practice observed on farms in the mountains is the use of containment barriers to prevent topsoil from being washed away. This is how Juan Antonio explains it:

> I position dry trunks in different places in the mountains so that the rainwater does not wash away the soil [...] I also plant pineapples. That's very good! When you plant them close to each other, they hold back all the leaf litter and topsoil. Barriers can also be made of stone or guano [dried royal palm leaves] so that – as long as it works – we won't have problems with erosion.

"On my farm we use organic fertilizer to increase production. The remaining manure from the cows and sheep is prepared to be applied to the plants, and so we do not need to use any chemical product", says José Luis, from Mango Bonito

(RBSR). Francisco, a little further south in the same Reserve, expresses himself in the same vein:

> we use animal manure to make biogas which is a very clean product and lets us save money on electricity and when we have to till the land, we use the traditional plough with oxen, so as not to impact the soil so much.

Learning new practices and obtaining new varieties

Farmers are receptive to innovation in both Reserves. The practices are not only from their own findings, but also the product of the dissemination of innovative practices by other farmers and learning through training programmes promoted by different institutions such as INIFAT (Instituto de Investigaciones Fundamentales en Agricultura Tropical Alejandro de Humboldt) and the Biosphere Reserves. In many cases, rural farmers have remarkable technical skills, due to their education, the training they receive as well as their direct and continuous social interaction. In both Reserves, the knowledge and practices of agroecological management are strengthened with the support of institutional actors. For example, insect traps are frequently used on farms to capture the individuals of the coffee berry borer and reporting farmers receive economic support from the State Plant Health Service of the Ministry of Agriculture of Cuba.

Yolanda, a farmer from La Tumba (RBSR), says that

> the exchange with other farmers has been very useful, because it has let me start applying some very important soil conservation practices on my farm. In addition, the training received from institutions like INIFAT has helped us increase our knowledge about very effective agroecological practices for pest and disease control for some crops.

An example is the use of green manure in their farms, as Segundo (RBSR) tells us: *"I planted canavalia that I was given as part of the COBARB project, and it really did the fallow land a lot of good. I'll continue to use it and keep the seeds"*.

Good agroecological practices would be of no use without adequate access to seeds, which is a key input. *In situ* conservation of seeds and farmer-to-farmer exchange networks provide security of access.

> Sometimes we have given maize seeds to other farmers to plant because theirs has got wet or doesn't grow [...] other times they have had to give some to me. This is how we can maintain the variety better, saving the grain and helping us [...], it is not often that a traditional farmer loses (his/her) seeds

states Víctor, a neighbour from La Munición (RBCT). Eduardo, from the same area, reports that

> thanks to the exchange mechanisms created between farmers and agricultural research institutions, we have received many varieties of seeds, for tomato,

lettuce, bell pepper, maize, cucumber and many others, and that has had a positive impact on the production and diversification on my farm.

Federico, also in the RBCT, states: "*I always try to save the good grains of traditional maize*".

Serguei, in RBSR, highlights the role of development projects in the diffusion of innovation: "the projects taking place in Sierra del Rosario have let us procure knowledge and seeds, which otherwise would have been very difficult for us to source". In addition to the traditional mechanisms for seed exchange, there are also agricultural diversity fairs organized by local governments with the support of the Biosphere Reserves and projects. These actions supporting genetic diversity add value to what already exists and, even though usually something new replaces the old, in many cases plant genetic resources accumulate and end up coexisting. Segundo, from the RBSR, explains this using the example of coffee:

Traditional coffee has been lost because other farmers have planted new varieties, but it is good to save seeds of traditional coffee, it has a better flavour, longer shelf life and good yield because it is well adapted to this zone. It has always worked well for my family. Traditional coffee promises to last for many years.

Genetic accumulation is also observed in other crops. This is what Diego, from La Munición (RBCT), Guantanamo, tells us: "I adopt what works for me, although I try to keep what my wife and I like. For example, producing various types of beans helps us vary our diet".

Dissemination of knowledge

Exchange is a very effective way for farmers to disseminate knowledge and innovation. "*When the farmer sees, he has faith*", is a popular saying. The farmer trusts his peers: "*I value sincere friendship. That's what I like most about these meetings*", said Eloísa (RBSR) during a COBARB Project training workshop. Not only knowledge but also genetic resources are exchanged during the fairs and workshops. Yolanda (RBSR) believes that exchange with other farmers and institutions changed her knowledge, practices and broadened her social relations in the region where she lives:

First of all, it has been excellent to be able to learn how much diversity there is both here and in the other areas that we visited. The exchange of experience, knowledge and seeds has helped us to improve our work. We have been able to identify our potential and that of the other farmers, with whom we have also been able to establish a rewarding friendship.

The workshops in which farmers are trained in agroecology, post-harvest work, or mushroom cultivation, have been advantageous for many farming families. "*I have applied all the techniques provided by the trainers on my farm and have appreciated the advantages they provide to the soil, and the products themselves"*, says Eduardo, in San Diego de Núñez (RBSR).

The experiences that farmers have had outside Cuba have also altered the local agricultural knowledge. In Chile, the participants had the opportunity of meeting farmers from the Chiloé archipelago, renowned worldwide for its diversity and agroecological production: *"There we learned the value of working together as a family"* said Eloísa (RBSR). Diego (RBCT) said that *"from that experience we came back with new lessons in pest management. We were very interested in how attracting beneficial insects can be achieved by planting flowers in the field and we are already doing this here"*. Crop pest control is an ongoing challenge for environmental and economic sustainability. Juan Bocourt, from the RBSR, is aware of this and notes that *"all the technical instructions and cultivation manuals that the COBARB project has given me have been invaluable in the elimination and control of the pests that affected my crops"*.

On the other hand, Juan Carlos affirms that

thanks to meetings with farmers from other reserves, I have learned new ways to give more value to the products I get from my land. Both myself and my family have learned that agroecological practices are the only thing that can guarantee that quality agriculture lasts over time.

Yolanda (RBSR) adds:

From the identified threats and risks for the productive system of our farm, we have a much clearer idea when it comes to prioritizing actions that we must take. It's our intention – mine and my husband's – to continue applying good conservation practices and contribute to extending their use.

Undoubtedly, the socialization of knowledge among farmers in the Reserves, as well as the interaction with experts and scientists, provides feedback that has yielded results. Eloísa (RBSR) sums this up as follows: *"I have been nourished by valuable experiences and I have been able to understand the value of my farm more and see more clearly all that I have in my environment and attribute more value to what I have"*.

The commercialization of agricultural products

Farmers in Cuba – either in cooperatives or not – have the possibility of contracting the commercialization of their products through the state-owned company Unión Nacional de Acopio. Farmers in both Reserves depend on the proper functioning of this system, especially in the case of crops such as coffee that generate large yields. This is also the case for beans and coconuts in the Cuchillas del Toa Biosphere Reserve but still, marketing of some crops cultivated in small quantities is challenging.

Some initiatives are making headway, however, and notably with women leading the way.

> The juicer that I brought to my farm has let me offer a new option to the community, while at the same time providing an outlet for the produce of my farm and those of other farmers. This way we all win,
>
>   says Eloísa (RBSR)

Mayi, a woman from Carambola (RBSR), self-manages an important community cultural project in her backyard:

> When I came here, more than twenty years ago, this place was barren. But today that has all changed and now I receive visitors from many different places. At the same time, we have integrated into the dynamics of the region, promoting the retrieval of our culinary tradition and its consumption.

Likewise, Caridad, a farmer from Baracoa (RBCT), leads her own micro-industry and manages the sale of her agricultural products in the city of Baracoa, where tourism is increasing. At the same time, Caridad passes on her knowledge: *"I have followed courses that have let me pick up very useful knowledge for my micro industry. This knowledge is shared between the women in the community, so that it can be useful to them too"*.

For Serguei (RBSR): *"The farmers' fairs have given me the chance to make my products known and in this way. I've also increased sales"*. Undoubtedly, these are true incentives for the farmers by helping improve the management of the sale of their production and, ultimately, a means to improve their standard of living. Lele (RBSR) also highlights the importance of the organization of farmers' fairs for marketing, because *"it prevents products from spoiling, which used to happen quite frequently"*. Meanwhile, José Luis (RBSR) states: *"Every time I am invited to participate in a fair, I always sell all my products, it makes me really happy"*. A farmer like Segundo (RBSR), who has worked in the mountains for many years, confirms that the marketing of his products has improved a lot in recent times:

> Acopio [Unión Nacional de Acopio] is doing its part and normally comes with a truck to pick up the products contracted in advance. I take part of what remains to the fairs to supplement my family's income and prevent it from going to waste in the fields; the rest of my production is for my family's own consumption.

Sustainability of the production system in the face of the process of change

The diversity of agroecosystems is a strength. The abundance of agricultural products, together with their multiple use, guarantees the integrity of the agroecosystem

(Perfecto et al., 2009). Hurricanes Gustav and Ike had a strong structural impact on the agroecosystems of the Sierra del Rosario Reserve in 2008 and also on the economies of many farming families. Farmers recall recovering of that impact thanks to the support of national institutions, of scientists associated with the Biosphere Reserves and also of other farmers. It also proved to be a source of innovation. For example, Manuel – an experienced farmer in the RBSR – found a new livelihood in producing charcoal after these hurricanes struck. After the damage caused to his coffee plantation, he opted to produce charcoal from the wood available of the uprooted trees, while continuing other activities such as animal husbandry. Over time, coffee would take on a renewed importance in the region, thanks to public policies to promote its cultivation. Over a period of three years, the new plantations started producing and a new crop – robusta coffee (*Coffea canephora*) – was added. Although this species of coffee already existed in Cuba as a phytogenetic resource (Pérez et al., 2010), its cultivation in the mountains is relatively new and it is much more hardy than traditional coffee: *"It is a more resistant coffee and tolerates the sun better"*, so it offers an opportunity for the farmer.

Not only are hurricanes a challenge to agricultural systems, but also other extreme weather events such as droughts and heavy rains, which are more frequent due to climate change (Pachauri and Meyer, 2014) and cause significant damage and losses. As a result, farmers are increasingly interested in perennial crops, such as fruit trees: *"I want to plant fruit trees because they are a good alternative"*, says Rolando, in the southern part of the RBSR. Gustavo, in the same area, illustrates how a family farm can regenerate, increase diversity and recover environmental services. Coming from an agricultural region outside the Reserve, it was a challenge for him to settle in the new property: *"When I first arrived, I was shocked because what I had before was something else. It seemed as if I couldn't live here. This was a mountain, so to avoid thinking too much, I started to work"*. With sheer effort, his farm was filled with avocados, mangos, oranges, guavas, lemons and other fruit trees. The family built structures to raise animals such as pigs, sheep, chickens, ducks and turkeys. They searched the bush for bees and adapted them to the farm: *"in the dry season the bees stick to the mangos, to the watering troughs and even to the sink. They don't die of thirst, they know how to survive, they are very intelligent and noble because they don't sting!"*. Currently, Gustavo's family keeps a garden for their own needs where they grow pumpkin, plantain, beans, malanga, yucca and yams that are also used for animal feed, in addition to *palmiche* (the fruit of the royal palm). The neighbouring farmers – who keep large livestock – provide him with animal manure to prepare compost, which complements what his sheep produce and thus constituting an excellent resource to fertilize the soil. Tree planting for timber such as teak and cedar has increased diversity, providing new habitats for wildlife, especially birds: *"at the beginning there was little vegetation and it seems as if the birds didn't have many places to perch and nest but now, they are much more common"*, asserts Neri, Gustavo's wife. On the property, you can see todies, thrushes, doves, wood pigeons, woodpeckers, blackbirds, emeralds, and *tojositas*: *"The little tojosita doesn't do any harm at all and over there you'll see the carpenter. There are already two of them living here! I feel that the farm is now a better place to live"*.

References

Beckford, C. and Barker, D. (2007) 'The role and value of local knowledge in Jamaican agriculture: adaptation and scale in small-scale farming', *The Geographical Journal,* vol 173, no 2, pp118–128

González Álvarez, A., Sánchez, Y., Arzola, D., Zamora, J.L. and Hernández, F. (2016) 'Agrobiodiversidad en la Sierra del Rosario, Cuba: el café (*Coffea arabica* L.) y otras claves de su configuración', *Agrotecnia de Cuba,* vol 40, no 2, pp3–8

Meneses Cabrera, T. and Cardozo Cardona, J. (2014) 'La Etnografía: una posibilidad metodológica para la investigación en cibercultura', *Revista Encuentros,* vol 12, no 2, pp93–103

Moreno Fraginals, M. (2014) *El Ingenio: Complejo económico social cubano del azúcar,* Editorial Ciencias Sociales, La Habana, 203pp

Narayan, D. and Patsch, P. (2002) *Voices of the Poor: From Many Lands,* World Bank and Oxford University Press, Washington, DC530pp

Pachauri, P.K. and Meyer, L.A. (2014) *Climate Change 2014: Synthesis Report. Contribution of Working Groups I, II and III to the Fifth Assessment Report of the Intergovernmental Panel on Climate Change,* IPCC, Geneva, Switzerland, 151pp

Pérez, C.A., Bustamante, C.C., Viñals, C.R. and Rivera, C.R. (2010) 'La fertilización nitrogenada de *Coffea canephora* Pierre var. Robusta, en función del rendimiento y algunos indicadores químicos y microbiológicos de suelos Cambisoles de Cuba', *Cultivos Tropicales,* vol 31, no 3, pp66–74

Perfecto, I., Vandermeer, J. and Wright, A. (2009) *Nature's matrix, linking agriculture, conservation and food sovereignty,* Earthscan, London, 240pp

Annex

List of scientific names and common
names of the species cited in the book.

Scientific name	Common name Spanish (Cuba)	Common name English
Abelmoschus esculentus	Quimbombó	Okra
Acanthurus spp	Barbero	Barbero
Acuneanthus tinifolius	Adedica	Adedica
Adelia ricinella	Jía blanca	Wild lime
Allium ascalonicum	Cebolla multiplicadora	Shallot
Allium cepa	Cebolla	Onion
Allium porrum	Ajo puerro	Garlic leek
Allium sativum	Ajo	Garlic
Allium schoenoprasum	Cebollino	Chives
Allium sp.	Ajo de montaña	Mountain garlic
Allophylus cominia	Palo de caja	Palo de caja
Amaranthus minimus	Bledo	Amaranth
Amazona leucocephala	Loro cubano	Cuban parrot
Amyris balsamifera, A. elemifera	Cuaba blanca	Balsam torchwood and Sea torchwood
Anacardium occidentale	Marañón	Cashew tree
Anaea cubana	Abanico	Cuban leafwing
Ananas comosus	Piña	Pineapple
Anas platyrhynchos	Pato	Duck
Andira inermis	Yaba	Cabbage Bark Tree
Anguilla rostrata	Anguila	American eel
Annona cherimola	Chirimoya	Cherimoya
Annona reticulata	Guanábana	Soursop
Annona squamosa	Anón	Sweetsop tree
Anobium punctatum	Polilla	Common house borer
Anolis luteogularis, A. homolechis, A. allogus, A. sagrei, A. barbatus, A. vermiculatus	Camaleón	Lizard
Anolis rejectus	Lagartija de hierba	Santiago grass anole
Antillophis andreai	Jubo magdalena	Caraiba
Apis mellifera iberiensis	Abeja ibérica	Spanish bee
Arachis hypogaea	Maní	Peanut
Ardisia dentata	Ardisia	Ardisia

Arracacia xanthorrhiza	Afió	White carrot
Artibeus jamaicensis	Murciélago jamaiquino	Jamaican fruit bat
Artocarpus altilis	Árbol del pan/Mapén	Breadfruit
Astraptes habana	Mariposa	Frosty flasher
Atta insularis	Bibijagua	Leafcutter ant
Avicennia germinans	Mangle prieto	Black mangrove
Azadirachta indica	Nim	Neem
Bactris cubensis	Palma pajúa	Pajua palm
Balistes vetula	Ballesta reina	Queen triggerfish
Bambusa vulgaris	Caña brava	Common bamboo
Benincasa hispida	Calabaza china	Wax gourd
Beta vulgaris	Acelga	Beet
Bixa orellana	Bija	Bija
Brassica oleracea	Brócoli, Coliflor	Broccoli, Cauliflower
Broughtonia cubensis	Broughtonia cubana	Cuban epiphytic orchid
Bryophyllum pinnatum	Siempreviva	Leaf-of-life
Bursera simaruba	Júcaro/Almácigo	Turpentine tree
Caesalpinia violacea	Yarúa	Yarua
Cajanus cajan	Gandul	Pigeon pea
Calendula officinalis	Maravilla	Marigold
Calophyllum antillanum	Ocuje	Ocuje
Calophyllum pinetorum	Laurel de las Indias	West Indian laurel
Calypte helenae	Zunzuncito	Bee hummingbird
Campephilus principalis	Carpintero real	Ivory-billed woodpecker
Canavalia ensiformis	Canavalia	Jack bean
Canis lupus	Perro jíbaro	Jíbaro dog
Capromys pilorides	Jutía conga	Desmarest's hutia
Capsicum annuum	Pimiento	Pepper
Capsicum chinense	Ají cachucha	Habanero pepper
Capsicum frutescens	Ají guaguao	Wild pepper
Capsicum spp.	Ají	Chili
Caretta caretta	Caguama	Loggerhead
Carica papaya	Fruta bomba	Papaya
Casuarina equisetifolia	Casuarina	Beach sheoak
Cedrela odorata	Cedro	Cigar box cedar
Ceiba pentandra	Ceiba	Kapok
Cenchrus purpureum	Hierba elefante	Napier grass
Chelonia mydas	Tortuga verde	Green turtle
Chilabothrus angulifer	Majá de Santa María	Cuban tree boa
Chlorostilbon ricordii	Zunzún	Cuban emerald
Chondrohierax wilsonii	Gavilán sonso	Cuban kite
Chrysophyllum cainito	Caimito	Purple star apple
Cichorium intybus	Achicoria	Chicory
Cinnamomum cassia	Canela China	Chinese cinnamon
Cinnamomum triplinerve	Canela	Canela
Citrus aurantifolia	Limón criollo	Key lime
Citrus aurantium	Naranja agria	Bitter orange
Citrus jambhiri	Limón francés	Rough lemon
Citrus limetta	Lima dulce	Sweet lime
Citrus medica	Lima dulce	Citron
Citrus paradisi	Toronja	Grapefruit
Citrus reticulata	Mandarina	Mandarine

Citrus sinensis	Naranja dulce	Sweet orange
Cittarium pica	Cigua	West Indian top shell
Clarias gariepinus	Claria	African catfish
Clusia rosea	Copey	Pitch apple
Coccoloba diversifolia	Guayacanejo	Tietongue
Coccoloba retusa	Cocolobo	Cocolobo
Coccoloba uvifera	Uva caleta	Seagrape
Cocos nucifera	Coco	Coconut tree
Coffea arabica, Coffea canephora	Café	Coffee
Colocasia esculenta	Malanga, Taro	Taro
Columbina passerina	Paloma rabiche	Common ground dove
Conocarpus erectus	Yana	Buttonwood
Cordia gerascanthus	Varía	Varia
Costus pictus	Caña santa	Caña santa
Crescentia cujete	Güira	Calabash tree
Cubophis cantherigerus	Jubo	Cuban racer
Cucumis sativus	Pepino	Cucumber
Cucurbita moschata	Calabaza	Crookneck squash
Cucurbita pepo	Calabaza	Field pumpkin
Cupania juglandifolia	Cupania	Cupania
Curcuma longa	Cúrcuma	Turmeric
Cyclura nubila	Iguana cubana	Cuban iguana
Cylas formicarius	Tetuán del boniato	Sweet potato weevil
Cyperus esculentus	Chufa	Yellow nutsedge
Daucus carota	Zanahoria	Wild carrot
Dichrostachys cinerea	Marabú	Sickle bush
Dioscorea cayenensis	Ñame africano	Yellow Guinea yam
Diospyros crassinervis	Ébano	Boa-wood
Drypetes alba	Hueso	Hueso
Eichhornia crassipes	Jacinto de agua	Water hyacinth
Elanoides forficatus	Tijereta	Swallow-tailed kite
Eleutherodactylus zugi, E. seileenae, E. limbatus, E. symingtoni	Rana	Frog
Elodea canadensis and Egeria densa	Elodea	Canadian pondweed and Large-flowered waterweed
Epicrates angulifer	Majá de Santa María	Cuban boa
Epinephelus striatus	Cherna	Nassau grouper
Eretmochelys imbricata	Carey	Hawksbill sea turtle
Eriobotrya japonica	Níspero	Loquat
Erythrina poeppigiana	Bucare	Mountain immortelle
Eupatorium sp.	Rompezaragüey	Snakeroot
Falco columbarius	Esmerejón	Merlin
Falco peregrinus	Halcón peregrino	Peregrine falcon
Felix catus	Gato	Cat
Ficus aurea	Higuera	Florida strangler fig
Ficus crassinervia	Jagüey	Jaguey
Fragaria vesca	Fresa	Wild strawberry
Gallus gallus	Gallina	Red junglefowl
Gambeya albida	Caimito blanco	White star apple

Gliricidia sepium	Piñón florido	Quickstick
Gonzalagunia sagreana	Palo semillero	Palo semillero
Guarea guidonia	Requia	American muskwood
Guettarda valenzuelana	Cuero	Cuero
Gymnanthes lucida	Yaití	Oysterwood
Halodule wrightii	Alga bajú	Shoalweed
Halophila engelmannii	Hierba estrella	Engelmann´s seagrass
Harrisia taetra	Cactus	Cactus
Hedychium coronarium	Mariposa	White ginger lily
Herpestes auropuntatus	Mangosta	Small Indian mongoose
Herpestes javanicus	Meloncillo	Javan mongoose
Hydrilla verticillata	Tomillo de agua	Waterthyme
Hyparrhenia rufa	Jaraguá	Jaragua grass
Hypothenemus hampei	Broca del café	Coffee berry borer
Ipomoea batatas	Boniato	Sweet potato
Juglans jamaicensis	Nogal	West Indian walnut
Justicia pectoralis	Tilo	Carpenter bush
Koanophyllon villosum	Rompebatallas	Florida Keys umbrella thoroughwort
Lactuca sativa	Lechuga	Lettuce
Laguncularia racemosa	Patabán	White mangrove
Lasiurus pfeifferi	Murciélago	Pfeiffer's red bat
Leucaena leucocephala	Aroma blanca	Giant wattle
Lippia alba	Flor de España	Bushy lippia
Lobatus gigas	Cobo	Queen conch
Luffa acutangula	Chiquá	Chinese ochro
Lutjanus analis	Pargo criollo	Mutton snapper
Mammea americana	Mamey de Santo Domingo	Mammee apple
Mangifera indica	Mango	Mango
Manihot esculenta	Yuca	Cassava
Manilkara zapota	Zapote	Sapote
Maranta arundinacea	Sagú	Arrowroot
Mastichodendron foetidissimum	Jocuma	False-mastic tree
Matayba apetala	Matayba	Matayba
Melaleuca quinquenervia	Niaulí	Punk tree
Melanerpes superciliaris	Pájaro carpintero	West Indian woodpecker
Meleagris gallopavo	Pavo	Turkey
Melicoccus bijugatus	Mamoncillo	Genip
Melipona beecheii	Abeja de tierra	Maya stingless bee
Mesocrapomys melanurus	Jutía andaraz	Black-tailed hutia
Mimosa pudica	Dormidera	Shameplant
Mus musculus	Guayabito	House mouse
Musa paradisiaca	Plátano fruta	French plantain
Musa sp.	Plátano vianda, Plátano fruta	Plantain, Banana
Mycteroperca microlepis	Aguají	Gag grouper
Mysateles prehensilis	Jutía carabalí	Prehensile-tailed hutia
Nandopsis ramsdeni	Joturo	Joturo
Nandopsis tetracanthus	Biajaca	Cuban cichlid
Nasturtium officinale	Berro	Watercress

Natalus primus	Murciélago oreja de embudo	Cuban greater funnel-eared bat
Nicotiana tabacum	Tabaco	Tobacco
Ocimum basilicum	Albahaca	Basil
Orbicella annularis, O. faveolata, O. franksi and O. cavernosa	Coral estrella	Boulder star coral
Orthosiphon aristatus	Té de riñón	Java tea
Oryza sativa	Arroz	Rice
Osteopilus septentrionalis	Rana de árbol	Cuban tree frog
Ovis orientalis	Chivo	Wild sheep
Oxandra lanceolata	Yaya	Black lancewood
Pachychilus violaceus	Caracol marino	Jute snail (common name for Pachychilus)
Pandion haliaetus	Águila pescadora	Osprey
Panulirus argus	Langosta	Spiny lobster
Parides gundlachianus	Mariposa de Gundlach	Gundlach's swallowtail
Passiflora edulis	Pasiflora	Passionflower
Peltophryne fustiger	Rana	Western giant toad
Persea americana	Aguacate	Avocado
Phania matricarioides	Manzanilla	Manzanilla
Phaseolus lunatus	Frijol caballero, habas lima	Lima bean
Phaseolus vulgaris	Frijol común, habichuela corta	Common bean
Pheidole megacephala	Hormiga leona	Big-headed ant
Phyllonycteris poeyi	Murciélago de flor	Cuban flower bat
Phyllops falcatus	Murciélago cubano comedor de higo	Cuban fig-eating bat
Picramnia pentandra	Aguedita	Florida bitterbush
Pilea microphylla	Frescura	Rockweed
Pinus caribaea	Pino macho	Pitch pine
Pinus cubensis	Pino de Mayarí	Cuban pine
Piscidia havanensis	Guaná candelón	Guana candelon
Pistia stratiotes	Lechuguilla	Water lettuce
Pithecellobium dulce	Guamúchil	Madras thorn
Plantago lanceolata	Llantén	Narrow-leaf plantain
Polymita picta	Polimita	Cuban painted snail
Pouteria campechiana	Canistel	Eggfruit
Pouteria sapota	Mamey colorado, sapote	Mamey
Priotelus temnurus	Tocororo	Tocororo
Prunum humboldti	Caracol marino	Sea snail (common name for Prunum)
Prunum tacoensis	Caracol marino	Sea snail (common name for Prunum)
Prunus armeniaca	Albaricoque	Apricot
Prunus occidentalis	Ciruela	Western cherry laurel
Pseudolmedia spuria	Macagua	Bastard breadnut
Psidium guajava	Guayaba	Guava
Pterois antennata	Pez león	Lionfish
Pteronotus parnellii	Murciélago bigotudo	Parnell's mustached bat
Ptiloxena atroviolacea	Totí	Cuban blackbird
Quercus sagrana	Encina	Cuban oak

Raphanus sativus	Rábano	Radish
Rattus rattus	Rata	Black rat
Rhizophora mangle	Mangle rojo	Red mangrove
Roystonea regia	Palma real	Cuban royal palm
Saccharum officinarum	Caña de azúcar	Sugarcane
Salvia officinalis	Salvia de Castilla	Sage
Samanea saman	Tamarindo de Puerto Rico	Cow bean tree
Sapium jamaicense	Piniche	Piniche
Scleria sp.	Cortadera	Nutrush
Sechium edule	Chayote	Chayote fruit
Sesamum indicum	Ajonjolí	Sesame
Solanum americanum	Hierba mora	American nightshade
Solanum lycopersicum	Tomate	Tomato
Solanum melongena	Berenjena	Eggplant
Solanum nigrum	Hierba mora	Black nightshade
Solenodon cubanus	Almiquí	Cuban solenodon
Sparisoma spp., Scatus spp.	Pez loro	Parrotfish
Spathodea campanulata	Espatodea	African tulip tree
Sphaerodactylus dimorphicus	Salamanca	Yellow-tailed dwarf gecko
Sphaerodactylus ramsdeni	Salamanca	Ramsden's least gecko
Sphaerodactylus schwartzi	Salamanca de Guantánamo	Guantanamo collared sphaero
Sphaerodactylus siboney	Salamanca	Gecko
Sphaerodactylus torrei	Salamanca de rayas	Cuban broad-banded geckolet
Spinacia oleracea	Espinaca	Spinach
Stachytarpheta jamaicensis	Verbena azul	Light-blue snakeweed
Starnoenas cyanocephala	Paloma perdiz	Blue-headed quail-dove
Sterculia africana	Anacagüita	African star-chestnut
Suriana maritima	Cuabilla de costa	Bay cedar
Sus scrofa	Puerco jíbaro	Wild boar
Syringodium filiforme	Hierba de manatí	Manatee grass
Syzygium jambos	Jambolón	Rose apple
Syzygium malaccense	Pomarrosa	Malay apple
Tagetes erecta	Flor de muerto	Big marigold
Tagetes patula	Clavel	French marigold
Talipariti elatum	Majagua	Blue mahoe
Tamarindus indica	Tamarindo	Tamarind
Tectona grandis	Teca	Teak
Teretistris fernandinae	Chillina	Yellow-headed warbler
Terminalia catappa	Almendra	Tropical almond
Thalassia testudinum	Hierba de tortuga	Turtlegrass
Theobroma cacao	Cacao	Cocoa tree
Thrinax radiata	Yaraguana	Thatch palm
Todus multicolor	Cartacuba	Cuban tody
Tournefortia gnaphalodes	Tabaquillo	Bay lavender
Tradescantia zebrina	Cucaracha	Silver inch plant
Trema micrantha	Guasimilla	Jamaican nettle tree
Trichechus manatus manatus	Manatí	Manatee
Trichilia hirta	Cabo de hacha	Cabo de hacha

Trophis racemosa	Ramón de caballo	White ramoon
Tropidophis pilsbryi	Jugo de cuello blanco	Cuban white-necked dwarf boa
Turdus plumbeus	Zorzal	Red-legged thrush
Vachellia farnesiana	Aroma	Aroma
Vigna radiata	Frijol mungo	Mung bean
Vigna umbellata	Frijol mambí/Frijol diablito	Rice bean
Vigna unguiculata	Frijol caupí	Cowpea
Xanthium strumarium	Guisazo	Common cocklebur
Xanthosoma sagittifolium	Malanga de chopo	Aarrowleaf elephant's ear
Xanthosoma spp.	Malanga	Malanga
Xiphidiopicus percussus	Carpintero verde	Cuban green woodpecker
Zachrysia guanensis	Caracol de tierra	Cuban land snail
Zea mays	Maíz	Maize
Zenaida macroura	Torcaza	Mourning dove
Zingiber officinale	Jengibre	Ginger

Index

Printed in the United States
by Baker & Taylor Publisher Services